Lightweight Cryptographic Techniques and Cybersecurity Approaches

Edited by Srinivasan Ramakrishnan

Published in London, United Kingdom

Lightweight Cryptographic Techniques and Cybersecurity Approaches
http://dx.doi.org/10.5772/intechopen.97981
Edited by Srinivasan Ramakrishnan

Contributors
Babu M., Sathish Kumar G. A., Gutumurthy J., Josephin Shermila P., Mohammed Abutaha, Haneen Dweik, Sumita Majhi, Pinaki Mitra, Mohammed Saeed Jawad, Mohammed Hlayel, Valery Ivanovich Korzhik, Vladimir Starostin, Victor Yakovlev, Muaed Kabardov, Andrej Krasov, Sergey Adadurov, Eric Järpe, Quentin Gouchet, Aamal Hafsa, Jihene Malek, Mohsen Machhout

First published in London, United Kingdom, 2022 by IntechOpen
IntechOpen is the global imprint of INTECHOPEN LIMITED, registered in England and Wales, registration number: 11086078, 5 Princes Gate Court, London, SW7 2QJ, United Kingdom

British Library Cataloguing-in-Publication Data
A catalogue record for this book is available from the British Library

Additional hard and PDF copies can be obtained from orders@intechopen.com

Lightweight Cryptographic Techniques and Cybersecurity Approaches
Edited by Srinivasan Ramakrishnan
p. cm.
Print ISBN 978-1-80355-732-8
Online ISBN 978-1-80355-733-5
eBook (PDF) ISBN 978-1-80355-734-2

Meet the editor

S. Ramakrishnan obtained a BE in Electronics and Communication Engineering from Bharathidasan University, India, in 1998, and an ME in Communication Systems from Madurai Kamaraj University, India, in 2000. He received his Ph.D. in Information and Communication Engineering from Anna University, India, in 2007. He has twenty-two years of teaching experience and one year of industry experience. He is a professor and the head of the Department of Information Technology, Dr. Mahalingam College of Engineering and Technology, India. He is also an associate editor for IEEE Access and a reviewer, guest editor, and editorial board member of numerous scientific journals. Dr. Ramakrishnan has published 163 papers in international and national journals and conference proceedings. He has thirteen books to his credit. He has advised nine Ph.D. scholars. His areas of research include digital image processing, soft computing, human-computer interaction, wireless sensor networks, and cognitive radio.

Contents

Preface

Cryptographic techniques are important due to the increasing digitization of most real-world applications. Lightweight cryptographic techniques suitable for devices with low computational capabilities that require fast computation are emerging due to the development of the Internet of things (IoT), 5G networks, and cloud computing. Since most organizations are opting for e-business models, suitable cybersecurity approaches and policies depending upon the nature and volume of business are needed. This book includes seven chapters that discuss various cryptographic techniques suitable for software-defined networks in the Internet of Things (SDN-IoT), Internet of Multimedia Things (IoMT), and SMS4-BSK pipelined cryptosystems. In addition, the book discusses intelligent cybersecurity threat management in modern information systems, keyless encryption techniques, and the differences between encrypted and non-encrypted data.

Due to the emergence of 5G networks, a new kind of network architecture referred to as SDN-IoT has been developed. In Chapter 1, the authors discuss SDN and IoT as well as lightweight cryptography, including hardware, throughput, latency, power consumption, and memory management. It also examines the three layers of SND-IoT (control layer, infrastructure layer, and application layer) and their security requirements in terms of lightweight cryptographic techniques suitable for 5G networks.

Encryption of image and video that is multimedia data is one of the challenging tasks in IoT due to its low computing capability. In Chapter 2, the authors discuss the chaos-based image and video encryption in IoT and address various issues of the IoM. They present a literature review of AES and its variant-based encryption methods along with a discussion of texture encryption and PLIE methods.

Lightweight cryptography can be achieved by a few mechanisms, among which pipeline cryptosystems play an important role. SMS4-BSK is a kind of pipeline cryptosystem designed to provide faster encryption with the required security in lightweight devices. In Chapter 3, the authors discuss SMS4-BSK cryptosystems and their real-time applications.

Encryption of digital images is one of the important tasks in cryptography. In Chapter 4, the authors provide a comprehensive overview of image encryption. The keys used for encryption are generated using a secure hash function and encryption is carried out using the AES algorithm. Key sharing is done using RSA.

Cybersecurity threat management is an essential requirement for various organizations. The cybersecurity policies to protect a business and its data are dependent upon the nature and volume of the business. In Chapter 5, the authors present several modern threat management frameworks and policies. They discuss various types of malware threats such as adware, spyware, ransomware, backdoors, and logic booms. They also examine reconnaissance tools such as social engineering and DNS harvesting tools. Furthermore, they discuss cloud-based security services.

Keyless cryptography is one of the alternatives to public key cryptography. Decryption in keyless cryptography is carried out only when all the secret shares are brought back. In Chapter 6, the authors survey various keyless cryptography mechanisms and present a few important mechanisms chapter using illustrations. These mechanisms include Shamir's protocol, Dean and Goldsmith cryptosystem, and EVS Key scheme. The authors also proposed a novel key sharing protocol and presented simulation results.

Distinguishing encrypted from non-encrypted data is one of the important tasks in cybercrime investigation to help police to recover deleted data. In Chapter 7, the authors examine various techniques for distinguishing between encrypted and unencrypted data (stream data and static data). They present a chi-squared distribution-based approach as well as propose a changed-point statistics-based method that they experimentally analyze.

This book is useful for graduate engineering students and researchers in computer science and electrical engineering as well as practicing engineers in the field of information and cybersecurity.

I would like to express my sincere thanks to all the authors for their contributions and efforts to bring about this wonderful book. My earnest gratitude and appreciation go to the staff at IntechOpen, particularly Author Service Manager Ms. Martina Ivancic. I would like to express my heartfelt thanks to the management, secretary, and principal of my institute and my department's faculty members. Finally, dearest thanks to my family members, especially my cute daughter Abirami.

Dr. Srinivasan Ramakrishnan,
Professor and Head,
Department of Information Technology,
Dr. Mahalingam College of Engineering and Technology,
Pollachi, India

Section 1

Lightweight Cryptographic Techniques

Chapter 1

Lightweight Cryptographic Techniques in 5G Software-Defined Internet of Things Networking

Sumita Majhi and Pinaki Mitra

Abstract

Lightweight cryptography (LWC) is an area of cryptographic techniques with low computational complexity and resource requirements. There must be a reason for using it in Internet of Things (IoT) network with a strict resource constraints environment. The key features of a 5G network are low latency, high throughput, heterogeneous network architecture, and massive connectivity. A new area of network architecture called SDN-IoT comes into the picture to control and manage IoT devices in a network with low latency and high throughput. SDN helps to reprogram the network according to the application's requirements. Also, higher mobile applications lead to higher data growth. SDN helps to secure, manage, and control the huge data in the network. SDN-IoT architecture divides the network into three layers: The infrastructure layer, the control layer, and the service or application layer. In this chapter, we are focusing on the LWC algorithms from different perspectives so that they will fit into different layers of SDN-IoT network. We will discuss all the pros and cons of implementing LWC algorithms in hardware and software environments and also, the different layers of the SDN-IoT network. We also discuss SDN security architecture and different performance metrics for LWC algorithms.

Keywords: lightweight cryptography, SDN, IoT, 5G, SDN-IoT security, architecture, cryptographic algorithm

1. Introduction

Software-Defined Networking (SDN) is an intelligent architecture in networking. It decouples the control and data plane which helps to improve the network performance and make it scalable, secure, and programmable. Internet of Things (IoT) network embedded with sensors nodes, RFID tags, smart cards, low resource devices which can communicate and share huge data to provide services to the clients. It is too difficult to provide security to the IoT system in heterogeneous and large networks. To combine SDN and IoT in a single architecture as SDN-IoT, it can make the infrastructure plane controllable, smart, reliable, and scalable [1]. IoT has many applications in different areas such as smart cities, smart vehicular networks, and security surveillance. To make these secure, SDN plays a huge part since it controls the whole network from its control plane. IoT devices capture and store sensitive information which is a great concern to make the network and physical devices secure from the eavesdroppers. 5G is the latest communication technology that is famous for low latency, massive connectivity, high throughput, and heterogeneous nature. By making SDN-IoT architecture in 5G, it can be flexible, dynamic and helps to improve the bad scalability due to hardware differences in heterogeneous environments. To make these happen there are many security challenges that need to be taken care of. One of the major necessities of 5G is low latency which is a real challenge with a huge growing market. Many cryptographic algorithms exist, but due to their high time and space complexity requirements, it will be a good choice to avoid these in a fast communication system like 5G. Recently, a lightweight cryptographic algorithm (LWCA) is a new area of cryptography applied in 5G [2]. These algorithms do not require much space, and the time complexity is also low which makes this technique applicable in IoT networks where limited battery life and strict physical constraints both need to be considered. Currently, many researchers tend to shift their focus from cryptographic aspects of security to lightweight security algorithms [3]. This helps the system to become less complex, provide high performance and also lower the cost. Traditional cryptographic techniques have high complexity as well as it is difficult to implement. In this chapter, we discuss different lightweight cryptographic algorithms and their applications in the SDN-5G network. There are three vulnerable areas in SDN-IoT architecture: Control Plane, Data Plane, and the Interfaces between Control and Data Plane due to their programmable nature and open access architecture.

LCAs are divided into four types: 1. Block cipher, 2. Stream cipher, 3. Hash functions, 4. Elliptic curve cryptography. Each of the techniques has its own strength and weaknesses. Based on the application's requirements, it can be used in different layers of SDN-IoT architecture. Block ciphers are AES, DES, DESL, DESX, DESLX, Piccolo, TEA, XTEA, mCRYPTON, PRESENT, TWINE, LBlock whereas SNOW-V and Espresso are the stream cipher presented in this chapter maintaining the strict requirements of LWCA. Apart from the above-mentioned algorithms also, GRAIN works as a stream cipher and can use as one of the LWC algorithms in the SDN-IoT network. It uses very few gates with high security and less power consumption. When it comes to security perspectives, HIGHT, ICEBERG, CLEFIA are good choices. Hummingbird holds both of the properties of block and stream cipher and can implement in both hardware as well software. This flexibility of implementation and maintaining the lightweight properties can use this technique in both the infrastructure and control layers of the SDN-IoT network in 5G. SHA-1, SHA-2, SHA-3, BLAKE2 are the algorithms used as Hash function techniques. We will discuss each of them and their applications in the SDN-IoT network later in this chapter. There are some application areas of cryptography where we work with such devices that operate on battery power and need cryptographic algorithms which consume less power, such devices are medical implant devices or environment-measuring devices. Although very few works have been done in this direction and many exposures are open for future research work in this subarea of cryptography which we mention here as lightweight cryptography (LWC). Most of the LWC algorithms can resist linear and differential attacks which are the basic criteria of any general cryptographic algorithm. It is important to measure the performance of LWC algorithms before using them in any application. One of the key criteria of the LWC algorithm is low latency. For that, some of the automobile sectors which require immediate response use LWC techniques for security purposes. IoT devices that require less CPU cost and memory consumption such as smart TV, tablet PCs are the application area of LWCAs. Also, medical sensors, smart agriculture sensors, RFID tag applications, electrical home appliances, automobile industry are the different applications of LWCAs.

2. Software-defined networking architecture

Software-defined networking (SDN), a network architecture, provides an environment that enables the network traffic and connectivity provided by a set of network resources that are centrally controlled or programmed by software applications. It manages the data traffic forwarding or processing functions, such as QoS, filtering, monitoring, or tapping. Except the traditional network architecture, the SDN architecture logically decouples data and control plane which helps to control the network centrally from the control plane. This helps to improve the scalability of SDN architecture. **Figure 1** illustrates the working principles of SDN. SDN architecture comprises three basic components: Data Plane, Control Plane, Application Plane.

2.1 Data plane

It consists of hosts and Open Flow (OF) switches. OF forwards data from source to a destination following the instructions of the Control layer. The data plane handles the data traffic forwarding and processing based on the configuration set by the control plane. It implements all the forwarding and processing decisions which have been made and commanded by the control plane. Also, the data plane needs to respond to network failure which is configured by the control plane. The controller plane interface with the data plane is called D-CPI. It executes the function which is capable of event notification.

2.2 Control plane

It controls the network resources of the data plane. Resources that are involved to forward and processing data traffic. It may comprise a set of SDN controllers based on the number of applications. The concern of control plane in case of multiple SDN controllers also to execute different applications with no overlapping with one another. To execute multiple application controllers, it has to communicate with one another.

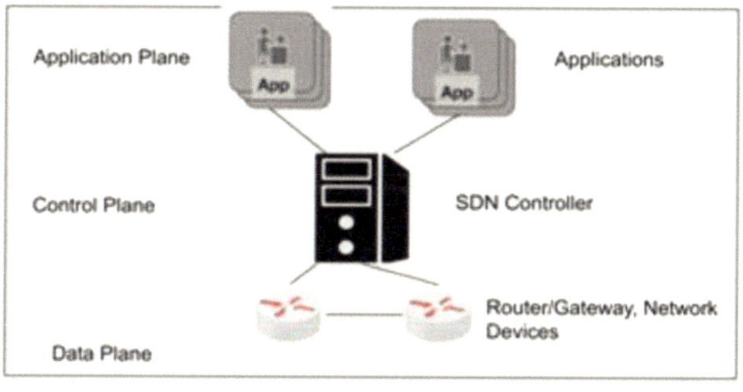

Figure 1.
Software defined networking architecture.

SDN controller which is a network operating system (NOS) is a logical concept. It is programmable and centrally keeps track of the global view of the network and data traffic. It dynamically configures the working strategy of the devices in the data plane.

2.3 Application plane

It consists of one or more applications that have the capability to communicate with them. The controller plane interface with the application plane is called A-CPI. Each SDN application may consist of a coordinator, SDN application logic, and A-CPI agent. An SDN application may invoke more than one SDN controller through the A-CPI interface to achieve its goal.

3. Internet of things

A heterogeneous network of smart devices that are connected and communicated for transferring a large amount of data to provide services according to client application's requirements technically called IoT. To make the physical structures, such as buildings, transportation vehicles, transportation networks, information technology networks more smart, secure, and automated, IoT is a great solution. It can make the physical devices interact with each other. It divides the whole network into three main abstract levels, sensing layer, network layer, the application layer. **Figure 2** illustrates the working principles of IoT with detailed descriptions of all layers.

3.1 Sensing layer

It is also called the perception layer. All the hardware integration has been done in this layer. In traditional internet, this layer is equivalent to the physical layer. It senses and collects data from physical devices. The data can be temperature, humidity, presence or absence of some observable, etc.

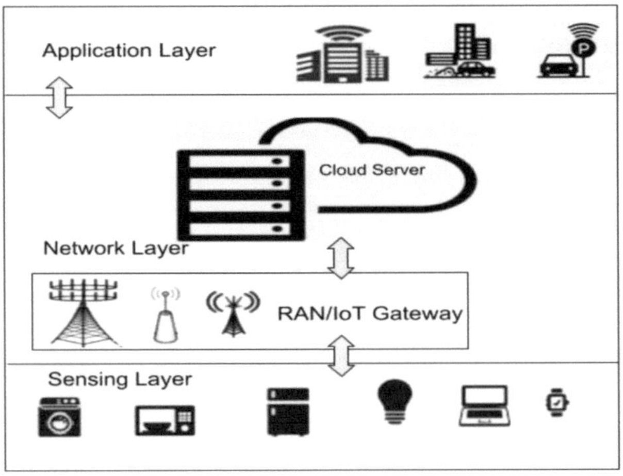

Figure 2.
Internet of things architecture in 5G network.

3.2 Network layer

It is also called the transport layer. It acts as a bridge between the sensing layer and the application layer. It transmits the data collected from the sensing layer and sends it to the application layer based on applications requirements. The network layer can be wired or wireless, sometimes both wired and wireless networks can together make an IoT network. This layer is responsible for receiving instructions from the controller.

3.3 Application layer

This is the last layer of abstraction in the IoT network. It receives data comes from the networks layer and based on this; it provides service. The applications can be a smart home, smart cities, smart vehicular network, security, and surveillance of a building.

4. Software-defined internet of things architecture

SDN architecture plays an important role for security purposes and traffic analysis. This architecture will help in SDN-IoT network for security purposes which is better than traditional internet in many different ways. As the name implies SDN working principle is based on software programming and control by the control plane. The advantages of an SDN network over the traditional network are it helps to modify the software functions based on dynamic requirements really quickly which is a great concern of traditional networks where the software functions are fixed. The inability for scaling, poor network security, and performance are the great concern in the traditional network which has been mitigated in SDN architecture.

Recently IoT is the fastest growing technology with dynamic characteristics. Traditional internet has static constraints which is the most difficult barrier to work with IoT networks. The dynamism and centralized control architecture of SDN helps IoT to be a great partner of SDN rather than of traditional internet. Also, SDN provides debugging tool which helps the IoT environment to enhance security which earlier was not possible on the traditional internet.

Figure 3 illustrates SDN with IoT architecture. This configuration divides the network into three layers, the infrastructure layer, control layer, and service layer or application layer.

The infrastructure layer is divided into two sublayers sensing layer and the communication layer. The sensing layer consists of different smart IoT devices and the communication layer consists of different data forwarding devices. The IoT devices help to create different IoT applications. These IoT devices collect a large volume of data from the physical plane. In comparison to SDN architecture, the infrastructure layer is equivalent to the data layer. The infrastructure layer consists of Open Flow (OF) switches which work exactly the same as what OF works in SDN architecture.

The next layer in SDN-IoT architecture is called the control layer which contains an SDN controller or a number of controllers. This layer is responsible for global controlling and monitoring communications between the infrastructure layer and the application layer. Multiple controllers can help in this regard if anyone fails since controllers are communicating with one another. This type of configuration called the multi-controller master–slave deployment model, where one controller act as a master controller and others are act as slaves waiting for the instructions of the master

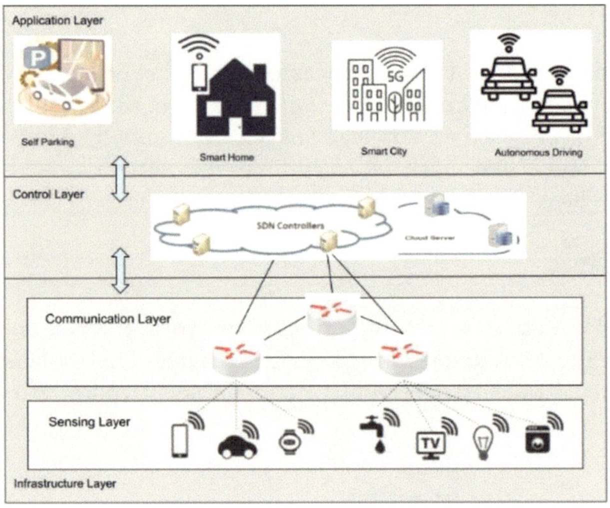

Figure 3.
Software-defined internet of things architecture.

controller if any one of the slave controllers fails to follow the instructions of the master controller. There are more than 30 controllers available up to now where some of them are open source and others are proprietary with their own programming languages and interfaces.

The application layer is responsible for different IoT services, such as smart home and smart city. This is also called the service layer in standard SDN-IoT architecture. The control layer communicates using the D-CPI interface to get required data from the sensing layer and using the A-CPI interface communicates with the application or service layer.

5. SDN security architecture

The difference between a traditional network and SDN is it decouples the control and data plane which divides the network into a set of components and interfaces. This unique feature of SDN makes it different from the traditional network also makes it vulnerable in terms of security. The controller plane is the central part of SDN, an attack in the controller plane may collapse the whole network. There are a few vulnerable areas that SDN encounters due to its centralized control and open programmable interfaces. The attacks can target to different areas of SDN described below:

Centralized Control: The policies defined by the application layer are followed by the control layer where the failure of the controller eventually leads to failure of the application program which may, in turn, fail the overall system.

Open programmable interfaces: There are three basic programmable interfaces that make SDN programmable in comparison to the traditional network.

1. **Between Application Layer and Control Layer:** The application plane submits policies to the control plane. Any malicious action while submitting the policy may fail the original application program to execute. REST API is used which

is a set of architectural constraints, not a protocol or standard, where it uses JSON, HTML, XLT, Python, or plain text. Since it is not a protocol like SOAP (Simple Object Access Protocol), rather a set of guidelines that can be implemented as per requirement, it takes less time to execute and is lightweight which is best for IoT networks.

2. **Between Control Layer and Infrastructure Layer:** Control layer submits the forwarding instruction to the infrastructure layer of network devices according to the application's requirement. Open Flow is a communication protocol used in this regard that is programmable and has centralized control.

3. **Between Different Controllers in the Control Layer:** Different controllers in the SDN-IoT structures may have been assigned for different tasks in the network. These are interconnected and dependent on each other. Failure of one controller invites failure not only for that particular task but also for the other task depending on the output of the task.

6. Lightweight cryptography

To measure the performance of lightweight algorithms, there are some performance metrics are described below.

1. **Hardware area:** This is the total area consumption to implement the code in hardware and is measured by gate equivalence (GE). In CMOS technology, the number of two-input NAND gate constitutes the total area referred to as GE and the unit used μm^2.

2. **Throughput:** This signifies the speed of an encryption or decryption operation that takes at a particular frequency and the unit used is Kbps. For a 5G network, the uplink and downlink speed is 10Gbps and 20Gbps, respectively. All of the encryption algorithms described below almost meet the criteria of 5G. The hardware and software implementation speed must be different for lightweight encryption algorithms. Some of the algorithms are generic to implement in both the hardware and software architecture but some of them strictly follow either hardware or software architecture. Based on their implementation strategy they can use different layers of the SDN-IoT network. One of the examples SIMECK 64/128 algorithm can implement in hardware and provide good GE and throughput value for using it in low-memory IoT devices, for that, it is suitable to use in the infrastructure layer of SDN-IoT network. On the other hand, the LED algorithm gives great throughput while implementing in a software environment as well as good security aspects which leads to the use of the algorithm in the cloud environment of the control layer. Throughput can be increased by using different signal processing methods, such as pipelining or parallel computations.

3. **Security:** The security level is defined by the key lengths in bits. The designer of the algorithm defined the security level based on the known attacks which have been applied to the algorithm and observed the behavior such as whether the algorithm can resist the attack or not.

4. **Latency:** Latency is defined as the total time required to complete a task. Here the number of clock cycles required for encryption of a single block message is defined as N. So, the latency is defined by:

Latency = N ∗ CriticalPathoftheCircuit

5. **Power and energy consumption:** Power consumption is dependent on the operating frequency, critical path, and so on the execution time. The unit used to define power consumption is μW. For hardware implementation power calculation is done by using GE and operating frequency. Energy consumption per bit can be calculated as and the unit used is μJ:

Energy = Latency ∗ Power/blocksize.

RAM/ROM memory: Total RAM or ROM memory required in bytes to execute the algorithm.

6. **Efficiency:** This is a trade-off between the performance of the algorithm and the cost to implement it. For hardware implementation, the efficiency can be calculated as:

Efficiency = Throughput/Complexity.

Complexity in terms of chip area is defined by GE and so the unit used to define complexity is KGE. For a software implementation, the efficiency can be calculated as:

Efficiency = Throughput/Codesize.

In this case, code size is defined in KB.

7. Lightweight cryptographic techniques in SDN-IoT for 5G network

The major constraint of IoT devices is it has limited resources in terms of processing power, storage and memory. This must be a primary reason when lightweight cryptographic (LWC) techniques came into the picture. This technique works in tight memory and resource constraints environments and has low computational complexity. Resources can be the size of the chip, cost of the IoT device, total speed, and power consumption. The size of the chip used in smart IoT devices must be small, so the encryption algorithm code size should be small enough to fit into these chips. The overall cost of the IoT devices should not increase much after using an encryption algorithm. The programming languages used to code the algorithm should be energy efficient, require less run time and memory [4]. LWC techniques are used for extremely low resource constraints devices which are communicating in IoT networks. LWC is one of the subbranch of cryptographic techniques. The battery technology is increasing relatively slowly and most of the encryption algorithm takes huge energy, so there is a trade-off between energy consumption and security.

5G mobile communication architecture is divided into three sections [2]—radio access network (RAN), core network, and application network. RAN in the infrastructure layer of the SDN-IoT network connects devices with one another to the control layer. In comparison with the hardware architecture, using software architecture reduces the equipment, development cost and improves flexibility. RAN can be modified to C-RAN which is a cloud/centralized RAN that works using software programming in 5G network. The security operations then moved to the cloud and implemented using software programs. This makes the cryptographic designer focuses more on the security aspects of the algorithm than the hardware efficiency which measures in GE. In this context, AES-256 is a great solution against quantum computing. There are a few algorithms discussed below, such as AES-256, SNOW-V, DES, Piccolo, Hash Algorithm, Espresso, which are potential for 5G security perspectives in the software platform. It is recommended to use in the cloud environment and is suitable for SDN-IoT architecture.

The security concern of SDN-IoT architecture comes into three layers as mentioned above. The first layer is an infrastructure layer, in IoT architecture, the infrastructure layer is equivalent to the sensing layer and in comparison to SDN architecture, this layer is equivalent to the data layer. All the LWC algorithms suitable for strict memory constraint IoT devices are used in this plane. The second layer is the control layer, which controls the overall system's architecture and the third layer is the service layer or application layer. Below we discuss all three layers and the potential LWC algorithms for each layer.

7.1 Control layer/infrastructure layer

SDN-IoT control layer is responsible for controlling the whole structure and traffic monitoring in a centralized manner, for that any malicious action can be detected from the control plane easily and for immediately taking an action. Multiple controllers connected by east/westbound interfaces with one another to maintain the connection. These interfaces are suffered from a lack of security support protocols and are easily vulnerable by the attacker. Identity-based cryptography (IBC) and elliptic curve cryptography (ECC) [5] are the two security solutions for this problem. Other cryptographic algorithms used in the control layer are Hash algorithms [6], AES [7], PRESENT [8], DES [9], etc.

1. Elliptic Curve Cryptography (ECC): ECC belongs to the category of LWC techniques. The key size has a significant role in cryptographic algorithms. The more the key size, the hard the algorithm to break. ECC which is based on a public-key cryptographic approach provides the same level of security as Rivest-Shamir-Adleman (RSA) algorithm but with a smaller key size. ECC with the 521-bit length of key provides the same level of security as Conventional RSA with a 15,360-bit length which implies ECC uses less memory than Conventional RSA signifies a great impact of mobile optimization. The key creation also takes less time in ECC which uses an elliptic curve to generate faster and smaller keys than the Conventional RSA algorithm which uses large prime numbers. To break the 228-bit ECC key, it would take more energy than the total energy required to boil all the water on the earth. This technique is called the next generation of cryptography since this is not a widely accepted method in the cryptographic system yet. It uses a complex mathematical algorithm to protect data which is a game-changer in the near future of cryptography. ECC is an asymmetric algorithm like RSA.

2. Hash Algorithms: Hash technique is considered a lightweight one-way authentication technique for generating a digital signature. SHA-1 and SHA-2 both are used in the control plane of SDN-IoT architecture of 5G network. The SHA-2 uses 256-bit digest whereas SHA-1 uses 160-bit digest confirms SHA-2 is more difficult to break than SHA-1 was the reason SHA-1 has not been used since 2010. The most recently developed Hash algorithm is SHA-3 which can be a

future concept of SDN-IoT security architecture. All the hash algorithms from the SHA family standardize by the National Institute of Standard and Technology (NIST). The total number of iterations taken by SHA algorithms is 80 which leads to a power-hungry situation. This in terns leads to a requirement of using the BLAKE2 hashing algorithm which uses eight rounds to generate a message digest of 256-bit. Also, the time and space requirement of the BLAKE2 algorithm is much better than SHA in digital signature-based authentication schemes. All the security threats can handle by SHA-2 algorithms up to today. SHA-3 will be used in the near future if any such situations are beyond the capability of SHA-2.

3. AES-256 [7]: It is a block cipher algorithm with a 128-bit data block and 256-bit key length used for encryption purposes in the control layer. The key length is variable; therefore, 128-bit key length requires 10 rounds, 192-bit key requires 12 rounds, 256-bit key requires 14 rounds. The throughput achieves for plaintext size 256 bytes is 22.67Gbps which is more than the targeted downlink speed requirement of 5G, which is 20Gbps. An AES implementation for RFID tag takes 3600 GE which is far beyond the minimum criteria of IoT nodes which is 2000 GE. This is one of the reason; AES-256 is used in cloud or software environments rather than hardware constraint environments such as the infrastructure layer of the SDN-IoT network. The control layer uses cloud and virtualization technologies to virtualize and centralize the function where securing the system using AES, DES, or other highly secure algorithms are recommended. Here security is more important than hardware cost. One of the famous lightweight IoT app "Flutter" includes AES-256 for encryption purposes.

4. SNOW-V: A 256-bit key length stream cipher can implement both in hardware as well software as SNOW-V [10]. Here V stands for virtualization. This algorithm takes the design and security techniques from SNOW 3G techniques. In this cryptographic technique, the throughput achieves for 256 bytes of plaintext is 26.37 Gbps. In comparison with SNOW 3G where the throughput achieves for 256 bytes of plaintext is 5.38 Gbps does not meet the minimum criteria of 5G network. So, there is a need to revise SNOW 3G to SNOW-V to meet 5G requirements. Both of the technique works with a key length of 256-bit. In a hardware implementation, it may require a large portion which may reach up to 19,179 GE. It is recommended to use this technique in a software environment and may not use for hardware constraint IoT devices.

5. DES: The main difference between AES and DES [9] is DES has a key size lesser than AES. This block cipher uses 56-bit keys with 64-bit blocks. Reducing the key size also reduces the hardware requirements of this algorithm while implementing it in a hardware environment. A smaller key size will lead to a lower security level. There are two variants of DES: DESL and DESX. Hardware implementation costs for DES and DESX are 2309GE and 2629GE. DESX uses the key-whitening technique to improve security performance. Another variation of the DES algorithm is DESXL which is the combination of DESL and DESX with GE 2169. On the other hand, DESL uses 1848 GE which is fairly a great deal for using it in hardware constraint devices such as RFID tags. DESL optionally uses the key-whitening method and avoid brute-force attack also reduces gate complexity by using serial hardware architecture and replacing 8 S-Boxes with a single box. This algorithm also improves the resistance against linear cryptanalysis and

differential cryptanalysis attacks. One of the variants of DES is 3DES also used in the control layer for encryption purposes.

6. Piccolo [11]: It is a 64-bit block cipher supporting 80-bit and 128-bit keys. From a security perspective as well as compact design aspects, this lightweight encryption technique can handle both. Both encryption and decryption take 818GE. This algorithm is not only famous for its minimal GE requirement but provides strong security against many attacks, such as Differential Attack, Linear Attack, Boomerang-Type Attacks, Impossible Differential Attack, Related-Key Differential Attacks, Meet-in-the-Middle Attack. This algorithm is suitable in a cloud environment which confirms to use it in the control layer of the SDN-IoT network. Due to its small memory requirements, it can also use in the infrastructure layer.

7. Espresso: This is a stream cipher that combines both of the primary constraints of 5G and IoT network which is hardware area requirements and throughput and provides a solution as an encryption technique. This technique is called the best trade-off between GE measure and throughput while security standards are also maintained. The hardware implementation requires 2045GE with 8.88Gbps throughput and 59 ns latency which meet most of the 5G requirements [12].

7.2 Infrastructure layer

The SDN-IoT infrastructure layer is integrated with constrained devices that require security algorithms that take less area to execute. Apart from timing, power, and energy [13] constraints, the area is another primary constraint of IoT devices. One of the metrics to measure the efficiency of the algorithm in terms of hardware area is gate equivalence. It is noted that GE less than 2000 is recommended for IoT devices in 5G network since they have very strict hardware and timing constraints, such as RFID tags, sensor nodes, smart and cards. All the encryption algorithms described here mostly have less than 2000 GE and are very efficient in terms of power consumption and timing.

1. TEA: TEA [14] is a lightweight cryptographic block cipher algorithm that is the fastest and is famous for its simple implementation. This algorithm is implemented in software with very few lines of code that can be implemented in any programming language is the main reason for its high-speed nature. This approach is resistant to differential attacks which is one of the major problems for IoT devices. XTEA and XXTEA are the two variants of TEA that are more efficient in terms of security and implementation. TEA follows the architecture of IDEA, a symmetric key block cipher that brings the gap between AES and DES. TEA uses 64-bit blocks and 128-bit keys. XXTEA which is the modified variant of block TEA, another variant of TEA works on variable-length blocks.

2. mCRYPTON: Another block cipher for resource constraint tiny devices, such as low-cost RFID tags and sensors is Miniature CRYPTON (mCRYPTON) [15]. mCRYPTON follows the architecture of CRYPTON, a 64-bit block cipher with key size options 64, 96, and 128-bits. mCRYPTON provides an economic hardware cost of 2400 GE for encryption under 0.13 m CMOS technology. This hardware cost is affordable for RFID tags and sensor nodes. For further size

reduction of 30% requires compact implementation of each component in both hardware and software.

3. PRESENT: PRESENT is another cryptographic technique known to be ultra-lightweight block cipher. In the case of PRESENT, it uses 1570 GE which is considered to be one of the lowest areas consumed while evaluating code for the algorithm. It is designed to be implemented in hardware and it is very difficult to implement it in software for the use of bitwise permutation.

4. HIGHT: HIGHT [16] is another lightweight algorithm for low resource IoT devices that require 3000 GE. In comparison with AES which requires 3400 GE, this algorithm takes less time to execute with a block length of 64-bit and a key length of 128-bit. Although the GE requirement of HIGH is much higher than other lightweight algorithms but the security aspects confirm this algorithm to use for IoT devices.

5. TWINE: TWINE [17] is another approach for lightweight cryptography with 1800 GE. It can be implemented in hardware as well as software signifies a good balance for hardware and software. TWINE is a 64-bit block cipher that supports the key value of 80-bit and 128-bit.

6. LBlock: LBlock [18] is a lightweight block cipher of block size 64-bit and the key size is 80-bit. The area efficiency of this algorithm is 1320 GE on 0.18 m technology with a throughput of 200 Kbps at 100 KHz and the software implementation on an 8-bit microcontroller requires 3955 clock cycles to encrypt a plaintext block.

7.3 Service layer or application layer

The application layer is responsible for different IoT services, such as smart home, smart city, etc. This is also called the service layer in standard SDN-IoT architecture. Different protocols work on the application layer of the IoT network. Message Queuing Telemetry Transport (MQTT) is one of the protocols used in the application layer which enhances machine-to-machine communication between client and server. The challenge of the MQTT protocol in 5G is to work with constraint IoT devices. The security improvement of the MQTT protocol is called Secure MQTT (SMQTT). The new version improves the security perspective of MQTT. For this purpose, there are many lightweight security algorithms are used, such as AES and RSA. Arduino is an open-source IoT development tool that uses the RSA algorithm. Diffie-Hellman (DH) and Elliptic Curve Cryptography (ECC) can be an alternative solution to the RSA algorithm. ECC is the most efficient public-key encryption technique in terms of power consumption for resource constraints IoT devices in comparison with other encryption techniques, such as RSA, Diffie-Hellman, and Digital Signature Algorithm (DSA). In ECC, it uses less key size and provides higher security. It is also a low latency algorithm that can be implemented in hardware as well as software environment leads to use this in infrastructure layer as well as cloud security environment. This technique also supports the minimum requirements of 5G security in terms of key value which must be at least 256-bit. We present the comparison of different LWCA in terms of GE, block length, and key length in **Table 1**.

SDN-IoT different layers	LWCA	Hardware area (GE)	Key length (bit)	Block length (bit)
Infrastructure layer	TEA [14]	2100	128	64
	mCRYPTON [15]	2400	64/96/128	64
	PRESENT [8]	1570	80/128	64
	HIGHT [16]	3000	128	64
	TWINE [17]	1800	80/128	64
	LBlock [18]	1320	80	64
Control layer and Infrastructure layer	AES-256 [7]	3600	128/192/256	128
	SNOW-V [10]	19,179	256	—
	DES [9]	2309	56	64
	Piccolo [11]	818	80/128	64
	Espresso [12]	2045	128	—
Service Layer	AES [7]	3600	128/192/256	128

Table 1.
Comparison of different LWCA for different layers of SDN-IoT network in 5G.

Experimental Study:

All the LWC algorithms presented in this chapter can implement in software/hardware/both environments. We are here to present a code snippet of the TEA algorithm using c language with a few lines of code. Here we present the algorithm for implementation purposes. The algorithm uses 32 rounds, although 16 rounds are sufficient. The term "delta" indicates here golden ratio serves for encryption/decryption purposes to get different values in each round.

Encryption:

1. Initialize:

    a. int. round = 0;

    b. unsigned long delta = 0x9e3779b9, a = 0, data[], key[], p, q;

2. p = data[0], q = data[1];

3. While (round <32)

    a. a = a + delta;

    b. p + = ((q < <4) + key[0])^(q + a)^((q> > 5) + key[1]);

    c. q + = ((p < <4) + key[2])^(p + a)^((p> > 5) + key[3]);

    d. round ++;

4. data[0] = p, data[1] = q;

Decryption:

1. Initialize:

a. int. round = 32;

b. unsigned long delta = 0x9e3779b9, data[], key[], p, q, a;

2. p = data[0], q = data[1], a = (delta<<5);

3. While (round >0)

a. q + = ((p < <4) + key[2])^(p + a)^((p> > 5) + key[3]);

b. p + = ((q < <4) + key[0])^(q + a)^((q> > 5) + key[1]);

c. a = a - delta;

d. round --;

4. data[0] = p, data[1] = q;

8. Conclusions

In the above, we describe different lightweight cryptographic algorithms that are used in different layers of the SDN-IoT network depending on the feature of the algorithm. We mainly focus on all the algorithms that satisfy the minimum requirements of 5G and IoT nodes in terms of throughput, power consumption, and hardware area requirement. There are many algorithms considered as lightweight encryption techniques based on different criteria, such as GE measure, code size, RAM/ROM, used. We here only focus on those encryption algorithms that satisfy the criteria of 5G network that is throughput value must be fair enough to support the minimum requirement of uplink and downlink speed and GE must be small enough for IoT nodes. Apart from throughput and GE constraints, there are other aspects that we have considered, such as implementation aspects of the algorithm, best suited in hardware or software environments. There is always a trade-off among different performance metrics of LWC algorithms. Depending on the requirements of the SDN-IoT network layer, the algorithms are set for a particular layer.

Abbreviations

LWC	Lightweight Cryptography
IoT	Internet of Things
SDN	Software-defined Networking
LWCA	Lightweight Cryptographic Algorithm
OF	Open Flow

NOS	Network Operating System
A-CPI	Application and Controller Plane Interface
D-CPI	Data and controller Plane Interface
SDN-IoT	Software-defined Internet of Things
SOAP	Simple Object Access Protocol
REST API	Representational State Transfer Application Programming Interface
GE	Gate Equivalence
RAN	Radio Access Network
C-RAN	Cloud/Centralized RAN
IBC	Identity Based Cryptography
ECC	Elliptic Curve Cryptography
RSA	Rivest-Shamir-Adleman
NIST	National Institute of Standard and Technology
AES	Advanced Encryption Standard
DES	Data Encryption Standard
SHA	Secure Hash Algorithm
RFID	Radio Frequency Identification
TEA	Tiny Encryption Algorithm
IDEA	International Data Encryption Algorithm
mCRYPTON	Miniature CRYPTON
MQTT	Message Queuing Telemetry Transport
SMQTT	Secure MQTT
DH	Diffie-Hellman
DSA	Digital Signature Algorithm

Author details

Sumita Majhi and Pinaki Mitra*
Indian Institute of Technology Guwahati, Guwahati, India

*Address all correspondence to: pinaki@iitg.ac.in

References

[1] Kanagavelu R, Aung KM. A survey on sdn based security in internet of things. In: Future of Information and Communication Conference. Cham: Springer; 2018. pp. 563-577

[2] Yang J, Johansson T. An overview of cryptographic primitives for possible use in 5G and beyond. Science China Information Sciences. 2020;**63**(12):1-22

[3] Biryukov A, Perrin LP. State of the art in lightweight symmetric cryptography. Cryptology ePrint Archive. 2017

[4] Pereira R, Couto M, Ribeiro F, Rua R, Cunha J, Fernandes JP, et al. Energy efficiency across programming languages: How do energy, time, and memory relate?. ln: Proceedings of the 10th ACM SIGPLAN International Conference on Software Language Engineering. 23 Oct 2017. pp. 256-267

[5] Shruti P, Chandraleka R. Elliptic curve cryptography security in the context of internet of things. International Journal of Scientific and Engineering Research. 2017;**8**(5):90-94

[6] Rao V, Prema KV. Comparative study of lightweight hashing functions for resource constrained devices of IoT. In: 2019 4th International Conference on Computational Systems and Information Technology for Sustainable Solution (CSITSS). Vol. 4. IEEE; 2019. pp. 1-5

[7] James M, Kumar DS. An implementation of modified lightweight advanced encryption standard in FPGA. Procedia Technology. (Elsevier). 1 Jan 2016;**25**:582-589

[8] Bogdanov A, Knudsen LR, Leander G, Paar C, Poschmann A, Robshaw MJ, et al. PRESENT: An ultra-lightweight block cipher. In: International Workshop on Cryptographic Hardware and Embedded Systems, 10. Berlin, Heidelberg: Springer; 2007. pp. 450-466

[9] Leander G, Paar C, Poschmann A, Schramm K. New lightweight DES variants. In: International Workshop on Fast Software Encryption, 26. Berlin, Heidelberg: Springer; 2007. pp. 196-210

[10] Ekdahl P, Johansson T, Maximov A, Yang J. A new SNOW stream cipher called SNOW-V. Cryptology ePrint Archive. 2018

[11] Shibutani K, Isobe T, Hiwatari H, Mitsuda A, Akishita T, Shirai T. Piccolo: An ultra-lightweight blockcipher. In: International Workshop on Cryptographic Hardware and Embedded Systems, 28. Berlin, Heidelberg: Springer; 2011. pp. 342-357

[12] Dubrova E, Hell M. Espresso: A stream cipher for 5G wireless communication systems. Cryptography and Communications. (Springer). 2017;**9**(2):273-289

[13] Mohd BJ, Hayajneh T. Lightweight block ciphers for IoT: Energy optimization and survivability techniques. IEEE Access. 2018 Jun;**18**(6):35966-35978

[14] Williams D. The tiny encryption algorithm (tea). Network Security. 2008 Apr;**26**:1-4

[15] Lim CH, Korkishko T. mCrypton–a lightweight block cipher for security of low-cost RFID tags and sensors. In: International Workshop on Information Security Applications, 22. Berlin, Heidelberg: Springer; 2005. pp. 243-258

[16] Hong D, Sung J, Hong S, Lim J, Lee S, Koo BS, et al. HIGHT: A new block cipher suitable for low-resource device. In: International Workshop on Cryptographic Hardware and Embedded Systems, 10. Berlin, Heidelberg: Springer; 2006. pp. 46-59

[17] Suzaki T, Minematsu K, Morioka S, Kobayashi E. Twine: A lightweight, versatile block cipher. In: ECRYPT Workshop on Lightweight Cryptography. 28 Nov 2011;**2011**

[18] Wu W, Zhang L. LBlock: A lightweight block cipher. In: International Conference on Applied Cryptography and Network Security, 7. Berlin, Heidelberg: Springer; 2011. pp. 327-344

Chapter 2

A Survey of Lightweight Image Encryption for IoT

Haneen Dweik and Mohammad Abutaha

Abstract

IoT networks serve as a way for various devices interconnected over the internet to exchange data with each other and with other services. Most smartphones, laptops, and other communication devices are connected to the cloud today, making data accessible to everyone. There are many applications for IoT, from smart IoT applications to industrial products. Encryption is one of the best ways to make IoT networks secure since so much data is being transferred. A lightweight block cipher is one of the most sophisticated means for overcoming the security problems inherent to IoT networks. Because of the limited resources available to nodes, classical cryptography methods are costly and inefficient. In this paper, we have compared the systems, we have found that these modifications were made to the original AES algorithm, while the original algorithm security remains robust, the modified AES algorithm remains lightweight and faster, providing more satisfaction for embedding in IoT devices and sensors that consume little power. Furthermore, this algorithm enhanced the AES-ECC hybrid encryption system, which has good flexibility and versatility, and optimized the design of the ECC function according to the characteristics of wireless sensor networks. Using Salsa20/12 stream cipher, the texture images can be encrypted using bit masking and permutation procedures and as part of a new scheme for encrypting 3D objects, which complements the existing methods for 3D object encryption. With PLIE implemented in Python, the encryption time was approximately 50% faster than that of AES using the throughput increase, faster encryption time, and minimal complexity.

Keywords: internet of things (IoT), lightweight, image, AES

1. Introduction

In the field of cutting-edge remote media communication, the Internet of Things (IoT) is quickly establishing itself as a new paradigm. In the Internet of Things, people, data, processes, and things are connected to make network connections that are more relevant and useful than ever before. With the rapid advancement of IoT, it is exposed to numerous risks and challenges, such as handling huge amounts of data, processing energy efficiently, responding to security threats, and encrypting/decrypting huge amounts of data. The concept refers to a system of interlinked computing items, such as RFID tags, sensors, actuators, and cell phones; digital machines; and people, allowing the sharing of data over a network without the need for human-to-human interactions. In an IoT world, massive amounts of raw data will be continuously collected, requiring real-time sensor data streams as well as techniques for converting this raw data into useful information. Furthermore, data privacy and security will be a serious concern. A cryptographic algorithm designed for a device with incredibly low resources will have different design criteria than one commonly used. Modern cryptography has evolved from this very specific area into lightweight cryptography. The low energy requirement of these algorithms makes them resistant to physical attacks.

The storage space requirements of multimedia applications are more challenging due to the size of multimedia data, the need for real-time processing, transmission delay, and security protection. Many new applications have emerged in the Internet of Things (IoT) and cloud computing fields, where multiple devices and servers perform thousands of operations at the same time. Multimedia applications require real-time processing, resulting in a critical role for encryption and decryption speed. A variety of technological fields, including smart cities and homes, have benefited from the Internet of Multimedia Things (IoMT). Most multimedia contents require large storage discs to be uploaded and streamed to different devices. As a result, video data or media content can be thoroughly analyzed if an issue occurs. IoT devices are low-powered and small in size, so they have been needed a cloud platform or third-party storage device to store, operate, and process information collected.

A majority of encryption is related to encrypting and decrypting text messages or documents, but images are also a prime bearer of crucial information, therefore, they have been needed to be encrypted. The encryption process involves modifying the pixels of an image so they had no longer representative of the original image. Once the receiver receives the encrypted image, it must be decrypted to reconstruct the image. Having encrypted images ensures that even if an interceptor gets access to a picture during transmission they had incomprehensible to them. Another practical use of encrypted images is for the security of biometric data. Fingerprint and retina scans involving biometric identification have become increasingly common, so these data must be securely shared and stored. When data is encrypted, it can be unintelligible to the intruder even if it is accessed maliciously.

IoT security has been emphasized by many organizations and research agencies. Open Web Application Security Project has identified privacy issues, inadequate authentication/authorization, lack of transport encryption, and poor physical layer security as the main causes of cyber-attacks on IoT. Identifying a device, validating its identity, authorizing it, establishing keys and managing them, as well as establishing trust and reputation are the five features in IoT security. Cryptographic primitives can help accomplish all of these objectives, including authentication, access control, non-repudiation, confidentiality, integrity, and availability. **Figure 1** shows the Thrust area in IoT security.

This chapter is organized in the following manner: in Section 2, I have presented a literature review. In Section 3, I have discussed the Advanced Encryption Standard (AES) and have described its detailed architecture framework. Section 4 presents the compared the current algorithms and approaches. Section 5 presents the results and discussion, while I have discussed the conclusions and provided further suggestions.

2. Literature review

Several systems and approaches have been proposed to address the challenges and restrictions involved with the encrypted transmission of big multimedia data.

Figure 1.
The thrust area in IoT security [1].

Moreover, the security of multimedia data needs to be researched further. This section presents studies that have been previously conducted in different categories.

In their study, Shadi Aljawarneh, Muneer Bani Yassein & We'am Adel Talafha [2], the encryption of big multimedia data, developed and designed a multithreaded encryption algorithm system. An advanced encryption standard (AES), genetic algorithms, and the Feistel Encryption Scheme (FEES) are used in this system. The system was evaluated concerning computational run time and throughput for the encryption and decryption process to analyze the performance of the system on actual medical data and benchmarked against the RC6, MARS, 3-DES, DES, and Blowfish algorithms. They have been also implemented the encryption system with a multithreaded programming approach to improve efficiency and performance. Finally, they have been tested their system against the sequential version to evaluate its resource efficiency. Comparing our system to other available encryption algorithms, their results showed that our system took the least amount of time to run and delivered a higher throughput. Furthermore, they have been also able to achieve a 75% improvement in computation run time and a 4-fold increase in throughput versus their sequentially structured version. Based on the security objectives, our algorithm performed better than existing algorithms in achieving the Avalanche Effect, and they could therefore include it in any encryption/decryption process of large, plain, multimedia data.

Haidar Raad Shakir [3], a new method that combines the Haar wavelet transformation with the Advanced Encryption Standard (AES), as well as pixel shuffling based on chaotic logistic maps. This method calculates the Haar wavelet transform from the original image and uses the fields of the approximation coefficient (LL) and detail confidences (LH, HL, and HH) to derive the different frequency domains of the image. Using the AES algorithm, the approximation part (LL) is encrypted, and the Haar wavelet transformation inverse is then applied. In addition to the chaotic logistic map, a shuffled image is used to impede malicious image reconstruction attempts to strengthen the encryption. Several representative methods from the literature were examined and compared to the method. According to the test results, it achieved better levels of encryption and less image degradation across a variety of images.

Yong Zhang [4], designed a C program that uses AES in cipher block chaining mode for image encryption. He presented an image cryptosystem that is compared with existing chaos-based image cryptosystems based on encryption/ decryption speed and security performance. In simulations, AES is shown to apply to image encryption, which argues against the commonly held perception that AES is not suited to image encryption. As a result of this paper, He recommends using AES-based image encryption as a benchmark for the speed of image encryption algorithms. And all other encryption algorithms whose speeds are lower should be discarded in practical communications.

Yong Zhang, Xueqian Li, Wengang Hou [5], According to their study, AES cannot cryptograph images in CBC mode. However, AES in CBC mode could be used to encrypt images. AES can be used to encrypt an image and generate an initial vector (IV). AES is secured by far, so the tested image cryptosystem is secure. Simulation results indicate the AES-based image cryptosystem is faster than some chaotic systems-based image cryptosystems. The tested system can thus be used as a reference for comparing other newly offered image cryptosystems. Cryptosystems for images that perform encryption and decryption slower than AES in the same computer need to be enhanced.

Sohel Rana. Saddam Hossain, Hasan Imam Shoun, Dr. Mohammod Abul Kashem [6], propose a lightweight cryptographic algorithm with 16.73% lower power usage than the existing cipher. Modern electronics and the internet will enable resource-constrained devices to become daily necessities for everyone, so data security will be an important consideration. Those devices will be communicating with one another incessantly, so information must be protected at all times. The implementation shows promising performance making the algorithm an ideal candidate for resource-constrained devices.

Charanjit Lal Chowdhary, Pushpam Virenbhai Patel, Krupal Jaysukhbhai Kathrotia, Muhammad Attique, Kumaresan Perumal, and Muhammad Fazal Ijaz [7], an analysis to decrypt and encrypt images using hybridization of Elliptic Curve Cryptography (ECC) and Hill Cipher (HC), ECC and AES (Advanced Encryption Standard), and ElGamal and Double Playfair Cipher (DPC). The measurements used in this analysis are (i) encode and decrypt times, (ii) entropy of the encrypted image, (iii) intensity loss of the decrypted image, (iv) Peak Signal to Noise Ratio (PSNR), (v) Number of Pixel Change Rate (NPCR), and (vi)Unified Average Changing Intensity (UACI). ECC and ElGamal cryptosystems offer asymmetric key cryptography, while HC, AES, and DPC provide symmetric key cryptography. Hybrid processes combine the speed and ease of implementation of symmetric algorithms with the security of asymmetric algorithms. According to the metric measurement with test cases, ECC and HC have a good overall solution for image encryption with smaller image sizes when using AES with ECC.

Bing Ji, LLijunWang an, d Qinghua Yang [8], an improved AES-ECC hybrid encryption system that has good flexibility and versatility and optimized ECC multiplication unit design according to the characteristics of wireless sensor networks. It was capable of generating and authenticating digital signatures at a faster rate. It also fully met wireless sensor networks' reliability, processing power, and power consumption requirements. AES encryption module is currently undergoing high-performance enhancements (increase throughput, decrease logic unit occupancy) and optimizations of ECC cryptographic module random point multiplications are currently being implemented. There are three properties of the scheme: (1) it provides better security with relatively low resource requirements, (2) it is straightforward to administer keys, and (3) it is resistant to some attacks and a digital signature can be generated and verified quickly and easily.

Alireza Jolfaei(B), Xin-Wen Wu, and Vallipuram Muthukkumarasamy [9], method encrypts texture images via bit masking and permutation procedures using Salsa20/12 stream cipher as part of a novel texture encryption scheme that complements the existing methods for 3D object encryption. As a result, the method has very low overhead and meets the security requirements, and protects the 3D surface geometry from partial disclosure by keeping the texture patterns hidden. Compared to full encryption and selective encryption (using the 4 most significant bits), the scheme has a higher speed-security profile. The schemes are implemented and tested with 500 sample texture images. Comparing the experimental results with full/selective encryption by 128-bit AES, the scheme demonstrated better encryption performance.

M. Sankari P. Ranjana [10], To protect the image data in the mobile cloud through privacy-preserve lightweight image encryption (PLIE), they have been introduced a method that keeps metadata on mobile while maintaining user privacy. Mobile data is split, distributed, and scrambled (SDS) to maintain user privacy and store it in the cloud. As a result, the throughput increases, the encryption time is sped-up, and the complexity is minimized. Using the PLIE method implemented in Python language, the encryption time was approximately 50% shorter than that of AES. They have been measured the performance of the existing method (AES) versus the method (PLIE) using various parameters. Furthermore, they have been evaluate the security level by presenting some security attacks.

3. Advanced encryption standard (AES)

The AES encryption algorithm is symmetric in the group, and there are three different key lengths: 128 bits, 196 bits, and 256 bits, with the packet size being 128 bits. The algorithm is reasonably flexible in its application. The AES algorithm is widely used in software and hardware. In the three key lengths, the 128bit key length is commonly used. The internal algorithm performs a ten-time iterative process when the key length is under. The five sections of the final round are joined by the Sub Bytes, S-box, Shift Rows, Mix Columns, and Add Round Key. AES has five different units of measurement: bits, bytes, characters, groups, states. A round of AES is composed of byte replacement (Sub Bytes), line displacement (Shift Rows), mixed column displacement (Mix Columns), key replacement (Add Round Key), and so on. AES algorithm design should meet three criteria during all phases of the data packet transformation, in the beginning, and ending stages of encryption:

1. Can resist all known attacks.

2. Fast and coding compaction.

3. Simple in design.

Figure 2 shows the process of AES Encryption and Decryption. It relies on the packet size and the length of the key, and it is controlled by the key. The iteration round of the number is controlled by the key and the length of the block.

As a **Figure 2**, a cryptographic algorithm is shown on the left and a cryptographic algorithm is shown on the right of the figure. A key expansion algorithm is shown in the middle of the figure. It consists of N iterations having four different steps: byte replacements (Sub Bytes), line displacements (Shift Rows), mixed column shifts (Mix

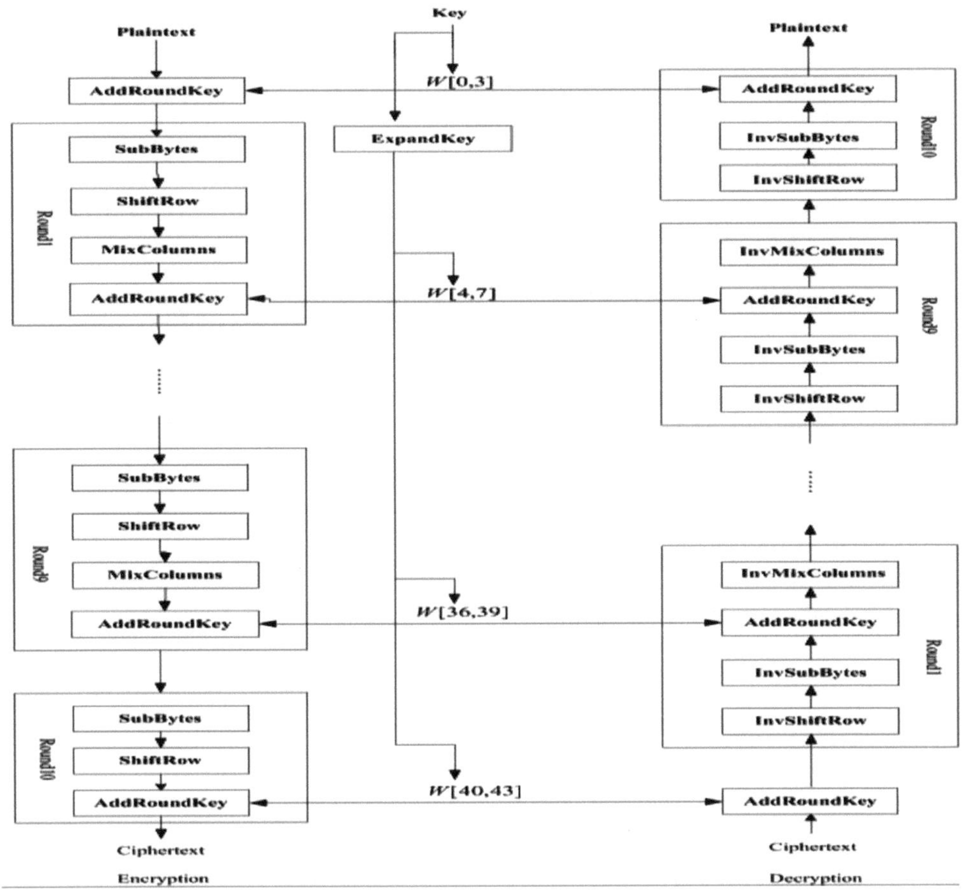

Figure 2.
Process of AES encryption and decryption.

Columns), and key shifts (Add Round Key). There are no mixed column transformations in the final round. The decryption algorithm is the opposite of encryption (inverse byte substitution, inverse shift rows, and inverse mix columns).

For full encryption, the data is passed through Nr rounds (Nr = 10, 12, 14). These rounds are governed by the following transformations:

- *Sub byte Transformation:* A non-linear substation table is used (s-box). It's constructed by multiplying inverse and affine transformation.

- *Shift rows transformation*: The offset of the left shift varies between one and three bytes, and the last three rows of the state are cyclically shifted.

- *Mix columns transformation*: The result is equivalent to multiplying columns of the states by a fixed matrix for each column vector. Note that the bytes are treated as polynomials rather than numbers.

- *Add round key transformation*: The round key is XORed with the working state, which is its overexpansion

- *key*: Even if an eavesdropper knows the plaintext and ciphciphertexte AES algorithm cannot be determined, because the secret key is known to both the sender and the receiver. According to its specifications, AES uses one of three key sizes (Nk). AES-126, AES-196, and AES-256 respectively use 128 bit (16 bytes, 4 words) and 196 bit (24 bytes, 6 words) key sizes. Key values have no weak point, unlike DES. All key values are equally secure, therefore no key-value renders encryption more vulnerable than the other. Key values are expanded via key expansion routines before being used in the AES algorithm. In addition to performing "on the fly" word expansion, this routine can be performed at any time.

3.1 AES key expansion

Add Round Key transformation uses a sub-key for every round, which corresponds to the number of bytes from the initial key. For example, AES-128 converts to 44 bytes per word, and every word are indexed as W[index] = [0 ... 43]. The first set of columns (W0, W1, W2, and W3) are all full with the given cipher key and the columns in locations that are multiples of four (W4, W8, W12, ... , W40) are all generated using the following three operations:

- Rot Word: Rot Word rotates a word to the left for one rotation.
- Sub Word: With Sub Word, individual bytes are replaced.
- Word Wi-4 and a defined constant from the Recon matrix are XOR'd with the result of Rot Word and Sub Word operations.

4. Compared the current algorithms and approaches

In refs. [8–10], by comparing three papers, I presented them in my study, which compared their results.

4.1 Hybrid algorithms

In ref. [8], three hybrid methods have been proposed for image encrypting and decrypting. This section describes the key generation, background process, and algorithm for image encryption and decryption. The algorithms are mentioned below:

- Elliptic Curve Cryptography (ECC) with Hill Cipher,
- ECC with AES,
- ElGamal with Double Play fair Cipher.

In this section I will explain two things:

1. How does the ECC generate the key?
2. ECC with AES for Image Encryption and Decryption.

4.1.1 Key generation of ECC

Here is an overview of the elliptical curve cryptography method. Elliptical curves are used over a finite prime field.

$$(F_p) = \{a, b, p, G\} \tag{1}$$

$$Y^2 \equiv x^3 + ax + b \ (\mathrm{mod}\ p) \text{ and } 4a^3 + 27b^2 \neq 0 (\mathrm{mod}\ p) \tag{2}$$

Where F_p is the finite field over a prime number p with generator G. a, b are curve parameters.

Whenever you multiply a point with different scalars, it creates every point on the curve. This is the generator of the curve G. To generate keys for elliptic curve cryptography, they have been defined the order of elliptic curve n as the smallest integer which when multiplied by generator G gives the zero point at infinity, that is, nG = O. **Figure 3** shows how does the ECC key generates.

4.1.2 ECC with AES for image encryption and decryption

For the encryption and decryption of images, Elliptic Curve Cryptography (ECC) is used in combination with AES. AES-128, AES-192, or AES-256 encryption and decryption are performed using the Cipher Feedback (CFB) mode. The initialization vector (IV) must be the same length and it should be a multiple of 16 or 24 or 32 bits, respectively. The initialization vector (IV) is not required for the Electronic Code Book (ECB) mode. AES encrypted bytes are converted to large integers to reduce operating costs by encrypting 2 × group size number of bytes in one ECC operation. This is because Base 256 represents values from 0 to 255 on the XY plane, and ECC encryption uses a point addition formula to encrypt any point on the XY plane. The algorithm is working with an 8-bit image whose pixel intensities range from 0 to 255. The benefit of performing such an operation is to eliminate the need to create a mapping table, which would otherwise be computationally impossible if an extremely large prime number was used to generate the finite field and share it between users.

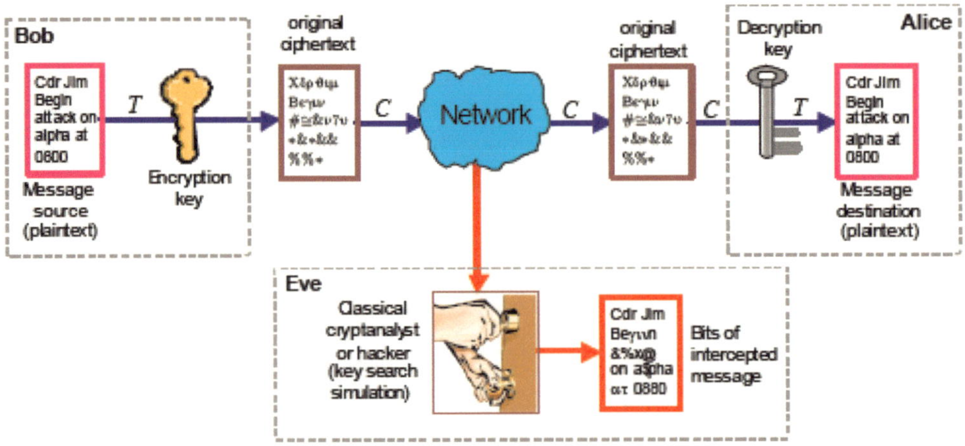

Figure 3.
Key generation of ECC [11]. Using ECC as an asymmetric approach, we have included key generation in their algorithm.

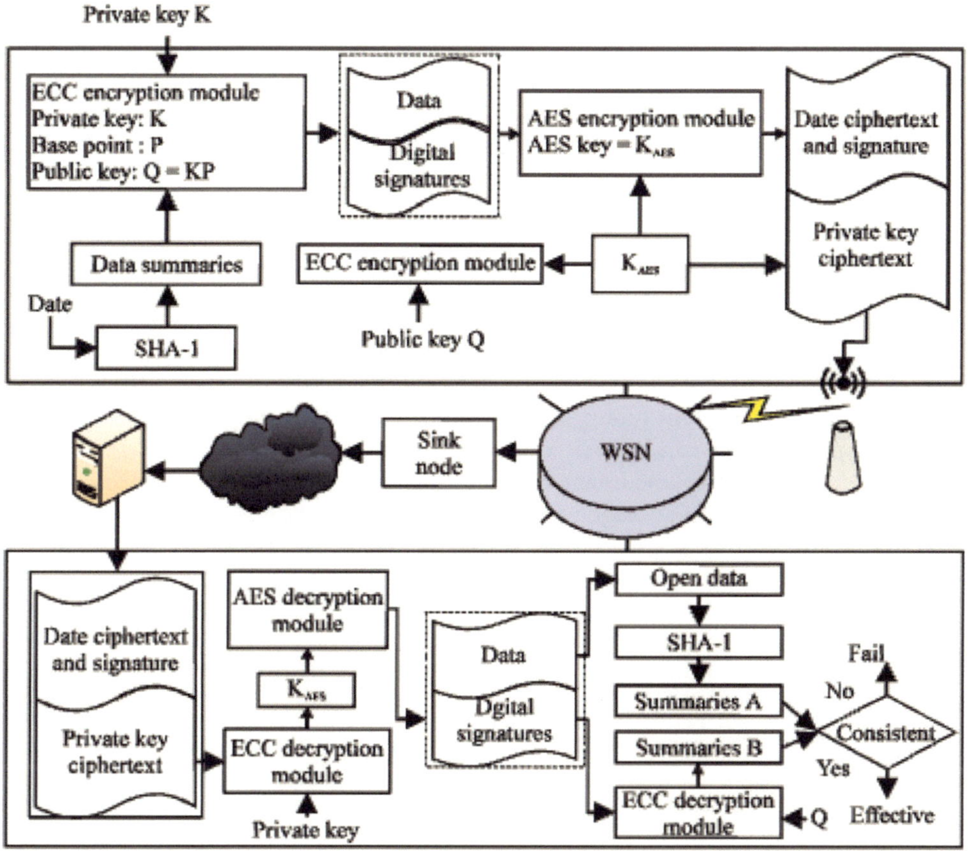

Figure 4.
AES-ECC hybrid encryption system [8].

Decrypting is achieved by reflecting the SSK coordinates for the x-axis and taking modulus p. The reflected point is then added to the x-axis using the point addition formula for performing the inverse operation. **Figure 4** shows an AES-ECC hybrid encryption system.

4.2 Texture encryption scheme

In ref. [9], Using Salsa20/12 for the upper nibble image encryption, the bitstreams of the lower nibble image are scrambled by permuting a zigzag pattern on the bitstream. They have been called our encryption mechanism 'Salsa Dance' since it is consistent with (Latin American) Salsa movement. Infer the steps of the encryption algorithm, P being the plain image, N being the nibble image, and C being the cipher image. If RGB is a 24-bit representation, each flat image, nibble image, or cipher image is represented by three M * N matrices, so R, G, and B color layers. For any pair of x ($1 \leq x \leq M$) and y ($1 \leq y \leq N$), multiply p (x, y) by n (x, y) and the result is the flat image, nibble image, or cipher image. P (x, y) and c (x, y) are the entry values for the plain-images, nibble images, and cipher images, respectively; n (x, y) $\in \{0, 1, \ldots, 15\}$.

A 24-bit texture image with one color layer has the encryption procedure described below. For the other color layers, the procedure is similar. By splitting every entry into upper and lower nibbles, we could break the plain image into two nibble images that correspond to x and y. For any x ($1 \leq x \leq M$) and y ($1 \leq y \leq N$), n1 (x, y) and n2 (x, y) are defined as follows:

$$n_1(x, y) = p(x, y) \bmod 2^4 \quad (3)$$

$$n_2(x, y) = (p(x, y) - n1(x, y)) \cdot 2^{-4} \quad (4)$$

Figure 5 shows a zigzag path for scanning an image with the dimensions 3 * 4 in **Figure 5a** if mod(s, 12) = 7. In this case, entry scanning starts at the 7th entry and ends at the 9th entry, which is the entry immediately before the initial one. Scanning involves the placement of bits, column by column, in a matrix sequentially as they have been encountered. Permutation affects both bit-plane image bits (diffusion) as well as the values of nibbles (confusion). A mod (s, 12) = 7, the permutation result of the test bit-plane image can be seen in **Figure 5b**. Following the permutation process, when the scrambled bit-plane image is combined with every 4 consecutive columns, the encrypted lower nibble-image with size M * N can be reconstructed.

The final step is to create the cipher image by combining the encrypted upper and lower nibble images. To summarize, the whole encryption process is as follows:

$$P = 2^4\, N_2 + N_1 \quad (5)$$

$$C = E(P) = 2^4 \cdot E_2(N_2) + E_1(N_1) \quad (6)$$

Where,

$$E_2(N_2) = \text{Salsa20/12}(N_2) \quad (7)$$

$$E_1(N_1) = \text{Perm}(N_1) \quad (8)$$

PK (plain image), N1, N2 (lower nibble image), and C (upper nibble image) refer to plain, lower, and upper nibble images, respectively. In decryption, cipher images are further divided into upper and lower nibble images. In 24-bit texture images, there is a close correlation between different layers of color in the image. The upper and lower nibbles are decrypted with the same keystream used in encryption, while the inner nibble is decrypted by inverse permutation. To meet this requirement, Salsa20/12 uses a 64-bit nonce each time the color layer is encrypted. The same message will

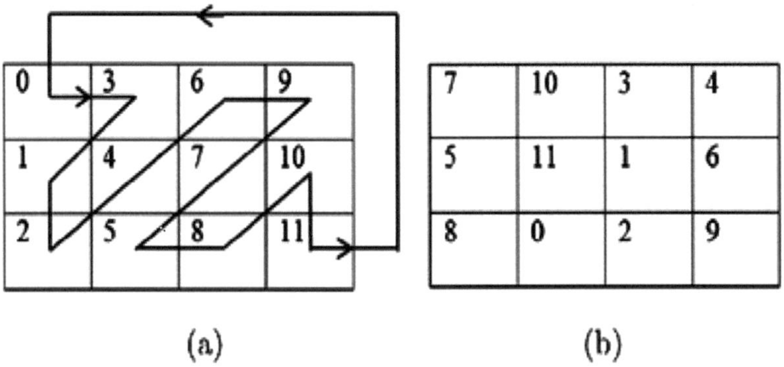

Figure 5.
(a) A zigzag path to scramble bits of a bit-plane image, and (b) permutation result [9].

never be encrypted twice in the same way so that there is always a different ciphertext. If the same nonce and key are used on two different plaintexts, then you can cancel the keystream out by masking the ciphertexts together.

4.3 PLIE method

In ref. [10], this method ensures image data security using three different processes, of splitting, distributing, and scrambling the images. In addition, it ensures user privacy by keeping metadata in the mobile device, and finally storing it on the cloud. A split image file is broken into two parts: the header and the contents. The header contains several privacy-protecting features, including the image type, size, date of creation, chunk size, height, width, and resolution. There are many chunks of content. For distribution, chunks may be divided based on a pattern, such as a key, a predefined function, or an individual chunk. PLIE categorizes patterns as odd chunks (file1) and even chunks (file2) that are sequentially repeated. The maximum number of chunks is m = (image size/chunk size)—header size, where the image size is the size in bytes of the image file, and chunk size is the size of the chunks.

5. Results

After studying the previous papers, I will be compared by showing them the different performance results. **Table 1** shows the different results.

Performance analysis	Size of input image encrypt	Size of output image decrypt	Algorithm used	Encryption time (seconds)	Decryption time (seconds)
In ref. [8], Eggs (Grayscale 256 × 256 Pixels) Image	256 × 256 pixels	256 × 256 pixels	AES-256 with ECC	2.82401	2.75127
In ref. [8], Eggs (colored 256 × 256 Pixels) Image	256 × 256 pixels	256 × 256 pixels	AES-256 with ECC	2.52632	2.50829
In ref. [8], Mona Lisa (Grayscale 256 × 256 Pixels) Image	256 × 256 pixels	256 × 256 pixels	AES-256 with ECC	2.84150	2.7866
In ref. [8], Mona Lisa (Colored 256 × 256 Pixels) Image	256 × 256 pixels	256 × 256 pixels	AES-256 with ECC	2.53642	2.52728
In ref. [9], Selective AES	M × 4 N	M × 4 N	ECB mode of AES-128	2.47	Not Calculated
In ref. [9], Full AES	M × 4 N	M × 4 N	ECB mode of AES-128	4.95	Not Calculated
In ref. [9], (Salsa Dance)	M × 4 N	M × 4 N	ECB mode of AES-128	1.00	Not Calculated

Performance analysis	Size of input image encrypt	Size of output image decrypt	Algorithm used	Encryption time (seconds)	Decryption time (seconds)
In ref. [10], Baby	256 × 256 pixels. That used file sized 4.9 KB	256 × 256 pixels	AES-128	0.0007	Not Calculated
In ref. [10], Leaf	File size 5.7 KB	256 × 256 pixels	AES-128	0.0009	Not Calculated
In ref. [10], Wheel	File size 6.5 KB	256 × 256 pixels	AES-128	0.00113	Not Calculated
In ref. [10], Ball	File size 8.7 KB	256 × 256 pixels	AES-128	0.00113	Not Calculated
In ref. [10], People	File size 13.3 KB	256 × 256 pixels	AES-128	0.0011	Not Calculated

Table 1.
Comparison of performance analysis.

5.1 Performance analysis

In reviewing the results in the table, we note that the time taken for coding in the first and second papers is greater than that spent on coding in the third paper.

5.2 Security

The following parameters were considered in the comparative analysis: ECC with AES, test samples [7].

- PSNR measures signal-to-noise ratio in decibels between two images. It is used to determine whether the original image is better than the compressed image. For the egg (grayscale), egg (colored), Mona Lisa (grayscale), and Mona Lisa (colored), the PSNR values are between 8 and 9.5.

- The NPCR metric value (%) is the expected change in the cipher image's pixels (when only one pixel of the plain image is changed) when the number of pixels from the input image is varied in the encrypted image. Based on the result of the Eggs (Grayscale), Eggs (Colored), Mona Lisa (Grayscale), and Mona Lisa (Colored) tests, it indicates that there is a significant number of pixels that differ from the original image in the encrypted image.

- The average value of the UACI (%) is 30% for varying numbers of pixels in the encrypted image from the input image. The UACI measure shows how secure an algorithm is against differential attacks, such as plaintext attacks or cipher-only attacks. Higher values indicate that this image is more resistant to such attacks. Values obtained for Eggs (Grayscale), Eggs (Colored), Mona Lisa (Grayscale), and Mona Lisa (Colored) range from 26 to 30%.

- Square error: in a decrypted image, the square error represents the discrepancy between the decrypted image pixels and the original pixels. For a good algorithm, the square error should be close to zero.

Evaluation metrics for	Mean squared error (MSE) in decrypted image	PSNR (dB)	NPCR (%)	UACI (%)
Eggs (grayscale)	0.0000	9.129	99.60632	28.90828
Eggs (colored)	(0, 0, 0)	(8.86691, 8.54116, 7.9478)	(99.63379, 99.64142, 99.6109)	(29.60341, 30.65091, 32.69575)
Mona Lisa (grayscale)	0.0000	8.62078	99.58191	30.37332
Mona Lisa (colored)	(0, 0, 0)	(9.39128, 8.98602, 8.7218)	(99.646, 99.63684, 99.62463)	(28.15297, 29.32043, 30.11092)

Table 2.
Metric measures.

Encryption schemes	Selective AES	Full AES	(Salsa Dance)
Between the original and encrypted image with the original key	6.0403 dB	6.0345 dB	6.0839 dB
Between the original and encrypted image with 1-bit different key	6.2709 dB	6.1846 dB	6.0821 dB
Between the encrypted images using the original and modified keys	8.5122 dB	8.3964 dB	7.7680 dB

Table 3.
Comparison of the PSNR values.

The PSNR, NPCR, UACI, and Mean Squared Error (MSE) metrics for Mona Lisa and Egg image are shown in **Table 2**.

Based on the test image in **Figure 7a**, **Table 3** offers the PSNR values for the encrypted images. Given the test image in **Figure 7a**, it is apparent that encryption using slightly different secret keys results in different Salsa Dance or AES cipher images. Salsa Dance, however, generates more dissimilar cipher-images than selective/full AES although the secret key is only changed by one bit. Thus, the method is highly sensitive to changes in the key, making the adversary's analysis of Salsa Dance even harder in terms of finding any relationship between the keys used.

5.3 Discussion and comparative analysis

According to the first paper [8], Compression of the image uses techniques that use less space to provide the same information, which solves the computation and high protection problem. The result is a low bandwidth, reduced storage space, and shortened computation times due to the compression.

According to the second paper [9], This paper describes a technical solution for meeting the confidentiality requirements associated with texture images that overcome the limitations of current techniques, in addition, large data volumes and high application requirements, including real-time performance, complexity, and security, are common.

According to the third paper [10], to reduce resource consumption, throughput, increase processing speed and reduce complexity, the PLIE method is an excellent choice for image encryption on mobile devices, It has been shown by a variety of

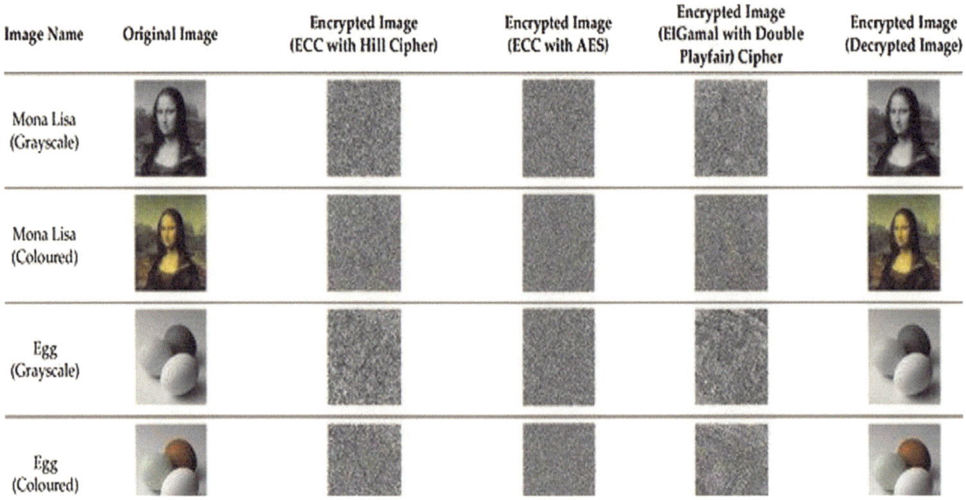

Figure 6.
Sample input and output for hybrid algorithms [8].

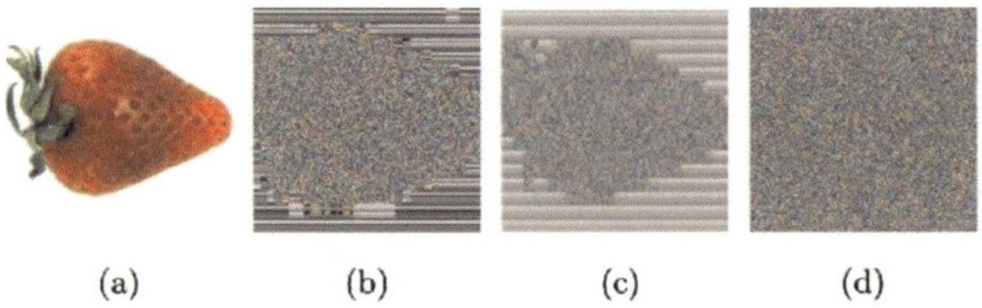

Figure 7.
Encryption results of a sample texture image: (a) original image, and (b) encrypted [9].

performance measurements to maintain privacy for users in mobile and to reduce encryption time by nearly 50% compared to existing methods such as AES.

For the study, input and output images include Mona Lisa (Grayscale 256 * 256 Pixels), Mona Lisa (Colored 256 * 256 Pixels), and Eggs (Grayscale 256 * 256 Pixels). Representative input and output images, with encryption and decryption algorithms, are provided [8]. **Figure 6** shows Sample input and output for hybrid algorithms.

An example texture image and its encryption results are shown in **Figure 7**. Salsa Dance seems to disrupt the correlation between entries of the image while both full and selective encryption using AES fail to destroy the coarse pattern.

6. Conclusion

Nowadays, all smartphones, laptops, and other communication devices connect to the cloud, making data accessible to everyone. IoT network is a group of various devices interconnected over the internet that exchange data between themselves and other services. It has a wide application range from smart applications to a variety of industrial applications. Encryption is one of the best techniques to guarantee end-to-end security in the IoT network, as the volume of data transferred is so high. Because nodes in an IoT network have limited resources, classical cryptography methods are costly and inefficient, so lightweight block ciphers are one of the most sophisticated ways to overcome security shortcomings in this environment. When we have compared the systems, we have found that these modifications were made to the original AES algorithm, while the original algorithm security remains robust, the modified AES algorithm remains lightweight and faster, providing more satisfaction for embedding in IoT devices and sensors that consume little power. Especially this algorithms that compared, improved AES-ECC hybrid encryption system that has good flexibility and versatility and optimized ECC multiplication unit design according to the characteristics of wireless sensor networks. It was capable of generating and authenticating digital signatures at a faster rate. It also fully met wireless sensor networks' reliability, processing power, and power consumption requirements. Salsa 20/12 method that encrypts texture images via bit masking and permutation procedures using Salsa20/12 stream cipher as part of a novel texture encryption scheme that complements the existing methods for 3D object encryption. Therefore, mobile data is split, distributed, and scrambled (SDS) to maintain user privacy and store it in the cloud. As a result, the throughput increases, the encryption time is sped-up, and the complexity is minimized. Using the PLIE method implemented in Python language, the encryption time was approximately 50% shorter than that of AES.

Author details

Haneen Dweik[1] and Mohammad Abutaha[2*]

1 Palestine Polytechnic University, Hebron, Palestine

2 College of Applied Professions, Palestine Polytechnic University (PPU), Hebron, Palestine

*Address all correspondence to: m_abutaha@ppu.edu

References

[1] PDFprof.com. CIAA Information Security Information Security PDF [online]. 2022. Available from: https://pdfprof.com/EN/PDF_Documents_Doc.php?q=3PDF48588-ciaa+information+security [Accessed: January 2, 2022]

[2] Aljawarneh S, Yassein M, Talafha W. A multithreaded programming approach for multimedia big data: Encryption system. Multimedia Tools and Applications. 2017;**77**(9):10997-11016

[3] Shakir H. An image encryption method based on selective AES coding of wavelet transform and chaotic pixel shuffling. Multimedia Tools and Applications. 2019;**78**(18):26073-26087

[4] Zhang Y. Test and verification of AES used for image encryption. 3D Research. 2018;**9**(1)

[5] Zhang Y, Li X, Hou W. A fast image encryption scheme based on AES. In: 2nd International Conference on Image, Vision, and Computing (ICIVC). New York: IEEE; 2017, 2017. pp. 624-628

[6] Rana S, Hossain S, Imam H, Mohammod D. An effective lightweight cryptographic algorithm to secure resource-constrained devices. International Journal of Advanced Computer Science and Applications. 2018;**9**(11)

[7] Chowdhary C, Patel P, Kathrotia K, Attique M, Perumal K, Ijaz M. Analytical study of hybrid techniques for image encryption and decryption. Sensors. 2020;**20**(18):5162

[8] Ji B, Wang L, Yang Q. New version of AES-ECC encryption system based on FPGA in WSNs. Journal of Software Engineering. 2014;**9**(1):87-95

[9] Jolfaei A, Wu X, Muthukkumarasamy V. A Secure Lightweight Texture Encryption Scheme. Image and Video Technology—PSIVT 2015 Workshops.2016. pp. 344-356

[10] Sankari M, Ranjana P. PLIE—A light-weight image encryption for data privacy in mobile cloud storage. International Journal of Engineering & Technology. 2018;7(4):368

[11] Rabah K. Theory and implementation of elliptic curve cryptography. Journal of Applied Sciences. 2005;**5**(4):604-633

Chapter 3

Perspective Chapter: The Importance of Pipeline in Modern Cryptosystem

Babu M., Sathish Kumar G.A., Gurumurthy J. and Josephine Shermila P.

Abstract

In this digital world, all the digital information transmitted through the wireless channel has a threat to its security. Along with security, encryption speed is also a significant factor in transmitting the data as fast as possible. The pipeline is the technique used to improve the throughput of the encryption process so that the amount of data encrypted per unit time will be increased. In this chapter, the design of the modern SMS4-BSK cryptosystem is briefed, various pipeline designs of SMS4 algorithms are surveyed and the pipeline implementation on SMS4-BSK cryptosystem is analyzed. The SMS4-BSK cryptosystem is robust, fast and has a throughput of 7.4 Gbps. This modern cryptosystem can resist all kinds of cryptanalysis attacks. The pipelining technique is implemented in this cryptosystem to improve the throughput further. The pipelining method is applied in the encryption architecture of the cryptosystem. The pipelined design is implemented in Kintex-7 FPGA. The design achieved a throughput of 9.9 Gbps. The pipeline implementation can be extended to the key scheduling architecture also as both the encryption and the key scheduling use the same architecture. As per the SMS4-BSK algorithm, the keys are generated in the host system to improve the throughput.

Keywords: pipeline, SMS4-BSK cryptosystem, throughput

1. Introduction

Pipelining is a very common and widely used technique to enhance the performance of a system without significantly investing in the hardware. In the pipelining technique, the computations are partitioned into a set of sub computations (elaborated in Section 2) and executed the sub computations in an overlapped fashion. The speed of execution is increased to an equal amount of sub computations obtained as a result of partitioning. The pipeline is used in different areas of computer design, such as memory access, instruction execution and arithmetic computation.

To improve the speed of computation, there is possible to adopt the following methods;

Method 1: Replicating the hardware.

Method 2: Partitioning the computations.

In the former method, by replicating the hardware, the performance can be improved by sacrificing the hardware cost. In the latter method, by partitioning the computation, the overlapped execution technique is implemented where the performance improvement is quite close to the former method.

Jin *et al.* have described [1] the pipelined design and folded design of SMS4 Block Cipher and implemented both the designs on the Xilinx Vertex-4 FPGA device. The implementation results show improved throughput in the former design and area coverage is minimized in the latter. According to the author, the proposed design might be the flexible choice for both the area-critical and the speed-critical cases. Gao *et al.* have proposed [2] rolling and unrolling architectures based SMS4 cryptosystem. The rolling structure uses a feedback system to control the entire processing mechanism. The unrolling structure is a fully pipelined architecture. The combination of rolling and unrolling provides good processing speed with an average clock cycle of one clock for processing128-bit. Han *et al.* have designed [3] an SMS4 architecture with optimization in power dissipation and implementation cost. The authors proposed a cryptographic algorithm for a programmable security processor. A three-stage pipelining and a 16-bit instruction set to enhance security are implemented in the design. The cost of the Processor and the code density of the design is significantly less. The round keys are stored in shared memory. A security scheme is proposed to protect these round keys. Zhao *et al.* have proposed [4] Galois Counter Mode (GCM) based SMS4 architecture. The architecture is fully pipelined to provide better performance. The structure can process 128-bit data on an average at each clock period. Full pipelined architecture is used here to improve the speed of encryption. The proposed design is implemented in both Vetex-4 and Vertex-5 FPGA. Zhao *et al.* have proposed [5] a novel implementation FPGA scheme for SMS4 cryptosystem. The throughput achieved in this scheme is 1.9 Gbps. Lee *et al.* have surveyed [6] the study pattern of computer architecture among students in implementing pipelining in the processor design. The design required only 21 Million Instructions per Second (MIPS). Abdel-Hafeez *et al.* have implemented pipelining [7] in Advanced Encryption Standard (AES). The architecture is implemented in Altera Max 3000A Field Programmable Gate Array (FPGA). The author claims that the pipelined AES design has a 16% higher throughput and 36% less hardware area than other designs. Guo *et al.* have proposed a pipelined AES [8] cryptosystem by combining pipelining with parallel processing and reconfiguration techniques. The design achieved a throughput of 8.83 Gbps. Teo *et al.* have implemented [9] pipelining in Data Encryption Standard (DES) cryptosystem. The architecture is implemented in Altera Complex Programmable Logic Devices (CPLD). Four stage pipeline approach is used to improve the throughput of the DES architecture. Babu *et al.* have implemented pipelining in SMS4 [10] cryptosystem. The design is implemented in Altera FPGA. The pipelining is applied to the Twisted Binary Decision Diagram (BDD) S-Box architecture. The Twisted BDD with m = 4 possesses good speed and throughput. The pipeline is implemented after the transformation block at each round. There are 32 round operations in the encryption architecture. Taherkhani *et al.* have designed [11] the pipelined DES cryptosystem and Vertex-6 FPGA to implement the architecture. The non-pipelined and the pipelined DES architectures are implemented in the same environment and analyzed. The results showed that the pipelined architecture's performance and throughput are better.

2. Overview of pipelined architecture

According to the eight great ideas of computer architecture [12], the pipeline is one of the techniques used to improve the Processor's performance.

Whenever a program is executed, it is executed through five phases;

a. Instruction / Opcode Fetch (IF)

b. Instruction Decode (ID)

c. Operand Fetch (OF)

d. Operand Execute (OE)

e. Operand Store (OS)

Figure 1 shows the organization of a Computer. Steps involved in the execution of a program are:

Step 1: The Instructions / Program as well as operands are stored in the Main Memory initially

Step 2: The Processor fetches the instruction from the Instruction Memory

Step 3: The Control Unit decodes the instruction and finds out the operand registers as well as the operation to be performed on the operand and sends the control signals to all the components

Step 4: Operands are fetched from the Data Memory

Step 5: Arithmetic and Logic Unit (ALU) executes the operands based on the operation decoded by the Control Unit

Step 6: The output from the ALU is stored back to the Data Memory

In the case of a non-pipelined architecture, after the ith instruction's execution, only the $(i + 1)^{th}$ instruction is initiated. After the $(i + 1)^{th}$ instruction's execution, only the $(i + 2)^{nd}$ instruction is initiated. The instructions will go on executing after finishing the previous instructions. This un-fashioned way of execution is known as instruction-wise interleaved execution or non-pipelined execution and it is shown in **Figure 2**.

tn: Execution duration of an instruction.

tp: Phase duration.

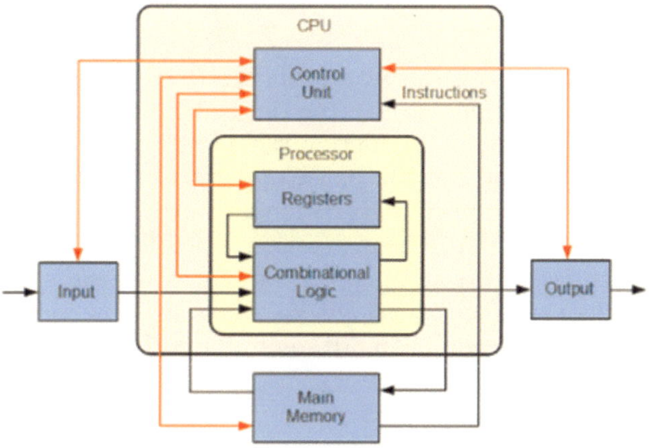

Figure 1.
Organization of a computer.

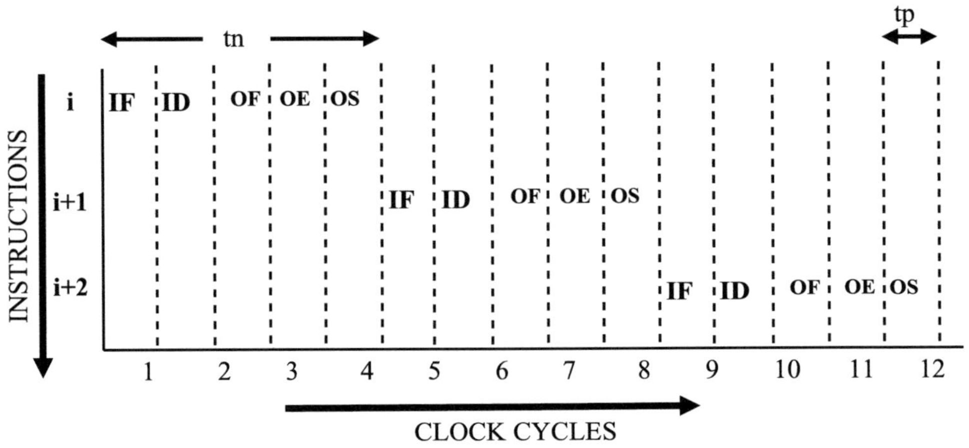

Figure 2.
Non-Pipelined Execution of instructions.

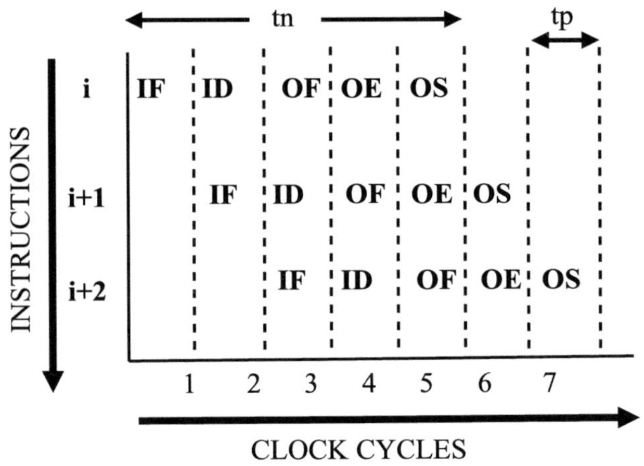

Figure 3.
Pipelined execution of instructions.

In the case of a Pipelined architecture, while the ith instruction is going from the first phase (IF) to the second phase (ID), the $(i + 1)^{th}$ instruction will be going to the first phase (IF). And while the ith instruction is going from the ID phase to OF phase, the $(i + 1)^{th}$ instruction will be going IF phase to ID phase and the $(i + 2)^{nd}$ instruction will be going to IF phase. This way of execution is known as phase-wise interleaved execution or pipelined execution. The instruction execution periods are overlapped and the parallel execution of the instructions is taking place and it is shown in **Figure 3**. No more than one instruction will be available in the same phase at each time slot. All the instructions will be in different phases. Phase-wise, there will not be any contention.

tp in non-pipelined architecture = 15.

tp in pipelined architecture = 7.

So, in the pipelined architecture, the instruction/program will be executed faster than the non-pipelined architecture.

The parameters used to evaluate the performance of the pipelined execution are:

Speed-Up ratio (S)

Frequency (f)

Efficiency (E)

Throughput (T)

Figure 4 shows the two-stage pipelined implementation. Consider a pipelined implementation with two stages S_i and S_{i+1}. The pipeline implementation is done by inserting buffers/latches between each stage. Input will go out of the latch when the clock is enabled.

τd = Latch delay (Delay taken by the input to go out of the latch).

τm = Maximum Phase Duration

$$Clock\ Period\ (\tau) = \tau m + \tau d \tag{1}$$

The clock period is the sum of the Maximum Phase Duration (τm) and the Latch delay (τd). All the phases may not have the same phase duration. So, the maximum phase duration among the five phases is considered for calculating the clock period.

In **Figure 5**, the number of instructions (n) = 5.

The number of clock cycles = 9.

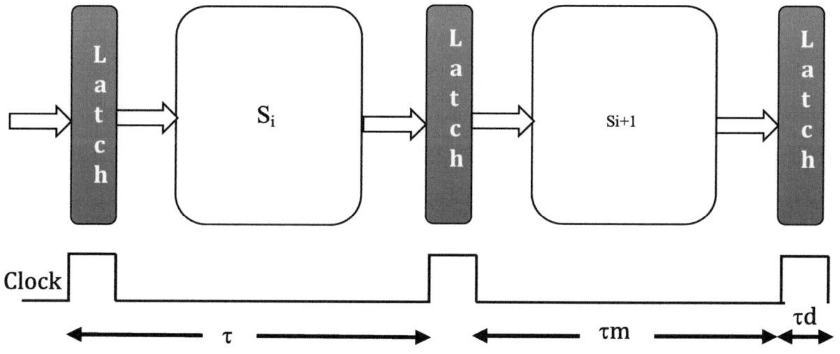

Figure 4.
Two-stage pipelined execution of instructions.

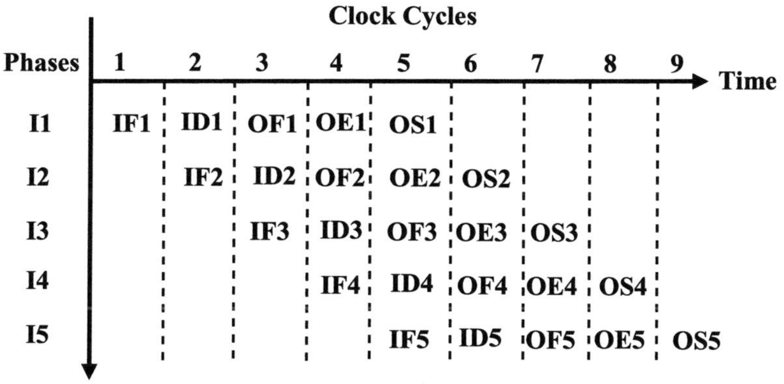

Figure 5.
Execution of instructions in a pipelined processor.

The number of Phases (k) = 5.
i.e.,

$$\text{The number of Clock Cycles} = k + n - 1 \tag{2}$$

Speed-Up ratio = Non-Pipelined Execution Time/Pipelined Execution Time.
i.e.,

$$S = \frac{n\, X\, tn}{(k + n - 1)\, X\, \tau} \tag{3}$$

If n> > k, k + n-1 ≈ n.
∵ tn = k x tp and tp = τ

$$S = \frac{n\, X\, k}{k + n - 1} \tag{4}$$

So,

$$S = \frac{tn}{\tau} \tag{5}$$

$$\text{Frequency}, f = \frac{1}{\tau} \tag{6}$$

$$\text{Pipeline Efficiency}, E = \frac{S}{k} \tag{7}$$

By substituting Eq. (5) to Eq. (7),

$$E = \frac{n}{k + n - 1} \tag{8}$$

$$\text{Throughput}, T = \frac{B\, X f}{N} \tag{9}$$

Where B is the message block size, f is the clock frequency and N is the adequate number of clock cycles utilized for the implementation.

3. SMS4-BSK cryptosystem

The SMS4-BSK cryptosystem [13] is a 128-bit symmetric key block cipher. The algorithm is designed to protect the message transmitted through the Wireless Local Area Network (WLAN). Thirty-two rounding operations are used in the encryption and the key generation algorithms. The encryption, as well as the key generation algorithms, use the same architecture. A unique non-linear S-Box, BSK Processing Block, is implemented in the design and operated over $GF(2^{16})$.

Figure 6 shows the process flow of the SMS4-BSK encryption architecture. The algorithm is designed especially for the sectors which use the shorter length messages (E.g., the Defense sector uses the shorter length messages to alert the enemy intrusion to the base army base camp).

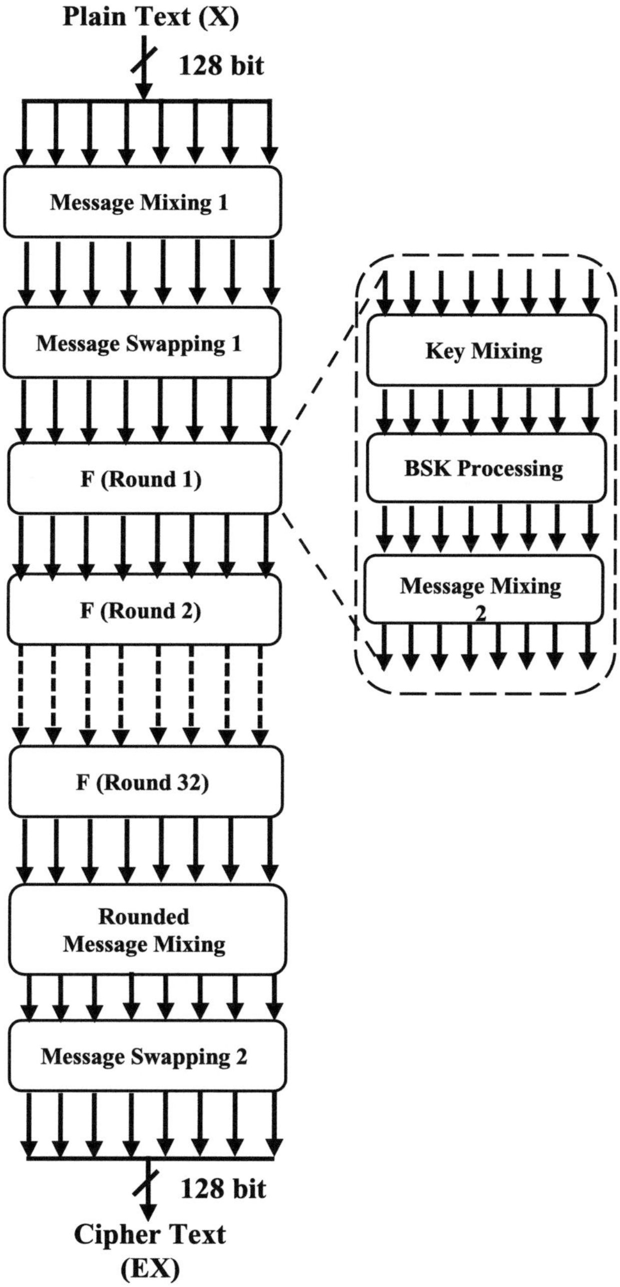

Figure 6.
Process flow of SMS4-BSK cryptosystem.

3.1 Encryption Algorithm

Step 1: Message Mixing - In this step, the plaintext is split into eight sub-blocks of equal size initially and then mixed with the nearest sub-blocks in a round fashion.

Step 2: Message Swapping – In this step, the first eight bits are swapped with the second eight bits in each sub-blocks.

Step 3: Key Mixing (Key generation is mentioned in the Key Scheduling Algorithm) – The generated half left and right key of 16-bit each is mixed with each sub-blocks after undergoing linear left/right shift.

Step 4: 32 Round BSK Processing – Step 10 of the Key scheduling algorithm is applied here.

Step 5: Message Mixing 2 – In this step, the processed message sub-blocks are linearly mixed with other sub-blocks.

Step 6: 32 Rounding

Step 7: Rounded Encrypted Message Mixing – In this step, the message sub-blocks are mixed in the opposite way as that of Step 1.

Step 8: Mixed Encrypted Message Swapping - In this step, the bits in the message sub-blocks are swapped in the same way as that of Step 2.

3.2 Key Scheduling Algorithm.

Step 1: Key Splitting 1 – The initial Key is split into two equal sub-keys.
Step 2: Key Splitting 2 – The even and the odd bits of each sub-keys are split.
Step 3: Key Mixing 1 – The odd key sets and the even key sets are mixed up
Step 4: Key Splitting 3 – Step 2 is repeated here
Step 5: Key Mixing 2 – Step 3 is repeated here
Step 6: 32 Round BSK Processing
Step 7: Cyclic Shifting
Step 8: BSK Processing – Step 6 and 7 are repeated here.
Step 9: Generation of Key 1
Step 10: Remaining Key Generation – Steps 6, 7, 8 & 9 are repeated to generate 32 keys.

The descriptions of both the Encryption as well as the Key Scheduling algorithm are elaborated in [13].

3.3 Key Features of SMS4-BSK Cryptosystem.

The SMS4-BSK design is more robust, i.e., the brute-force attack may take 2^{526} ns to decrypt the original message.

The cryptosystem is 1.1 times faster than the other SMS4 [14, 15] algorithms.

- The throughput achieved by the design is 7.4 Gbps.
- The power consumption of the hardware design is less.
- The cryptosystem has a keyspace of 2^{352}.
- The plaintext sensitivity is very close to 0.5.
- The key sensitivity is also very close to 0.5.
- The cryptosystem can resist known-plaintext attacks.
- The cryptosystem can resist chosen plaintext attacks.
- The cryptosystem can resist ciphertext-only attacks.
- The cryptosystem also can resist chosen-ciphertext attacks.

4. Pipelined SMS4-BSK cryptosystem

The pipelining is implemented in the encryption architecture of the SMS4-BSK cryptosystem. The pipelined encryption architecture of the SMS4-BSK cryptosystem is shown in **Figure 7**. A latch is introduced between every processing step. The processing delay of every step in the encryption architecture is different. So, the step with a higher processing delay may affect the encryption speed and throughput. By implementing a

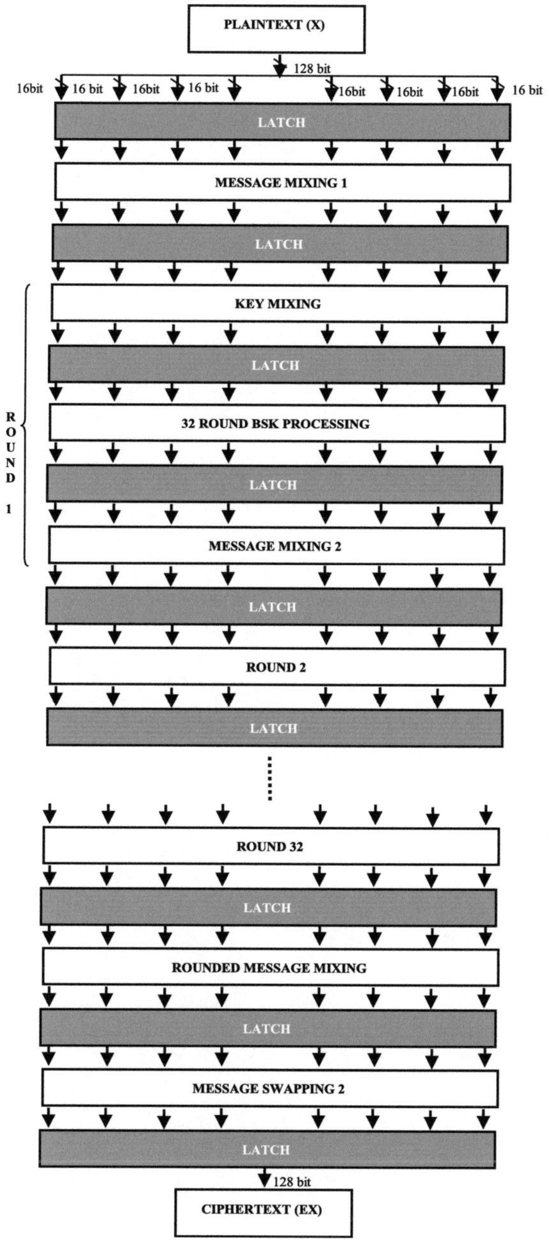

Figure 7.
Pipelined SMS4-BSK encryption architecture.

latch between every processing step, it will be possible to balance the data access rate at each step. The pipelined SMS4-BSK cryptosystem is developed in Xilinx Vivado 2017 Design Suite using Verification Logic Hardware Description Language (Verilog HDL) and implemented in Kintex-7 Field Programmable Gate Array (FPGA).

The Kintex-7 FPGA with the Device Name: XC7K410T and the Package: FBG676 is used to implement the cryptosystem. The number of Input and Output Buffers (IOB) available in the package are 400 and the requirement of the cryptosystem is 386. The synthesis report generated notes the timing summary and the area utilized by design. **Figure 8** shows the technology schematic of the pipelined SMS4-BSK obtained from the Xilinx Vivado 2017 Design Suite. The simulated timing diagram of the pipelined SMS4-BSK cryptosystem is shown in **Figure 9**.

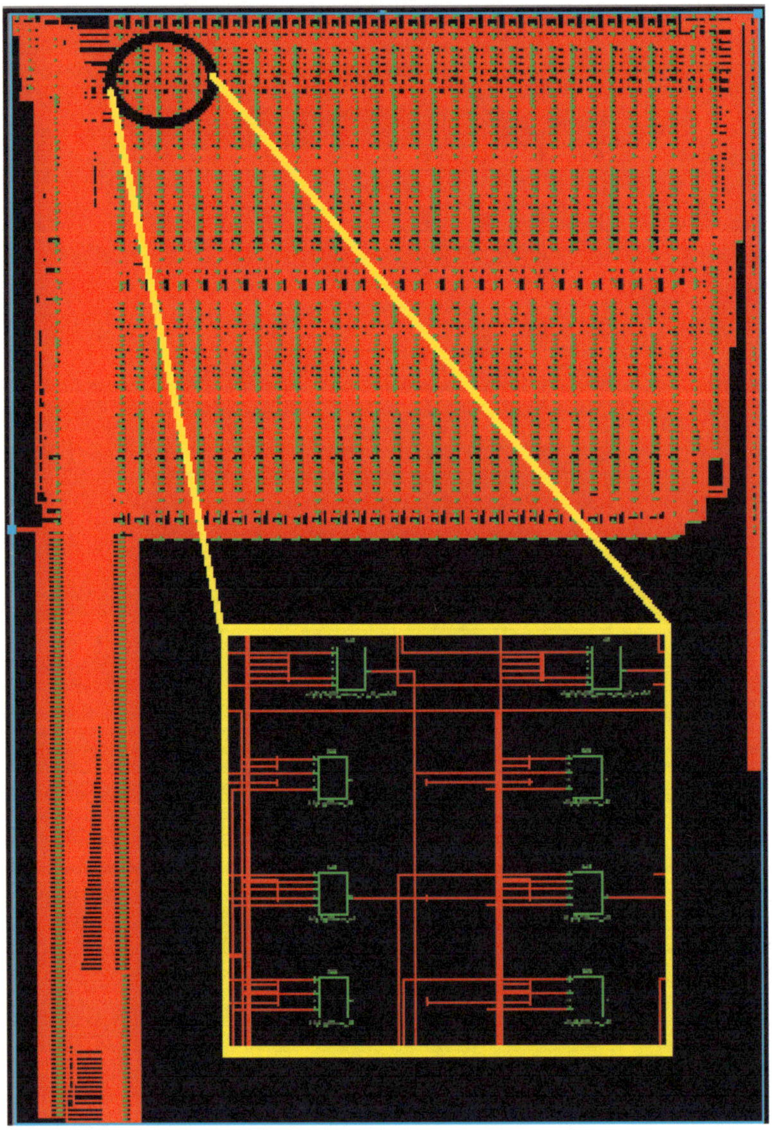

Figure 8.
Technology schematic of the pipelined SMS4-BSK encryption architecture.

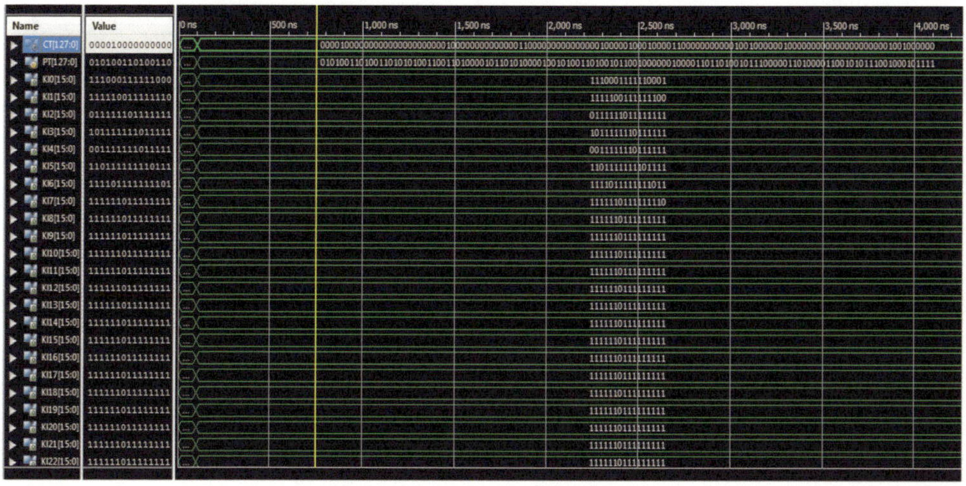

Figure 9.
Simulated timing diagram of the pipelined SMS4-BSK encryption architecture.

Architecture	Throughput
Twisted BDD S-Box SMS4	801 Mbps
Standard SMS4	1.9 Gbps
Folded & Reconfigurable SMS4	6.3 Gbps
SMS4-BSK	7.4 Gbps
Pipelined SMS4-BSK	**9.9 Gbps**

Table 1.
Comparison of pipelined SMS4-BSK with other SMS4 cryptosystems.

The throughput of the Pipelined SMS4-BSK algorithm is calculated using Eq. (10).

$$\text{Throughput} = \frac{B\,X f}{N} \tag{10}$$

Where B is the message block size (128 bit), f is the clock frequency (464 MHz) and N is the effective number of clock cycles utilized (6 Clock cycles).

$\therefore$ Throughput $\approx$ 9.9 Gbps.

The performance of the Pipelined SMS4-BSK encryption architecture is compared with the other SMS4 architecture and tabulated in **Table 1**. It is evident from **Table 1** that the throughput of the Pipelined SMS4-BSK cryptosystem is far better than other cryptosystems.

5. Conclusion

In this chapter, the implementation of the pipeline in the modern SMS4-BSK cryptosystem is discussed. The SMS4-BSK cryptosystem is more robust, has a large Keyspace, has optimum Key Sensitivity and Plaintext Sensitivity, is faster, has high throughput and can resist all the four major cryptanalysis attacks. The throughput of the SMS4-BSK cryptosystem is further improved to 9.9 Gbps by implementing a pipeline in the encryption architecture. The comparison of throughput of the various architectures of the SMS4 cryptosystem is shown in **Table 1**. It is evident from **Table 1** that the pipelined design of the SMS4-BSK cryptosystem has higher throughput. All the designs are implemented in Kintex-7 FPGA for comparison.

As a future enhancement, the pipeline implementation can be extended to the Key Scheduling architecture to improve the throughput further. A novel BM S-Box is being designed to replace the BSK processing block (Non-linear Transformation Block) of the SMS4-BSK cryptosystem to improve the speed and throughput.

Acknowledgements

We thank Ms. Saranya, S, UAT Analyst, AutoZone, U.S.A., for her support throughout the project. We extend our thanks to Ms. Preethy V for mentoring us in the project.

Conflict of interest

The authors declare no conflict of interest.

A. Appendix

A.1 Real-time applications of pipelined SMS4-BSKCryptosystem

Figure A1 shows the Inter-Vehicle Communication (IVC) implemented with the Pipelined SMS4-BSK cryptosystem. The communication between the vehicles is secured with the help of the proposed cryptosystem.

Figure A1.
Inter-vehicle communication with improved security.

Figure A2 explains a war field scenario: A short message transmission by a scout if an enemy intrusion is detected. The communication between the remote scout hub and the Army base camp is secured with the Pipelined SMS4-BSK cryptosystem.

Figure A3 shows the secured data transmission from the Smart Glucose Bottle to a caretaker or a nurse. The data sensed by the intelligent system is transmitted with the help of Internet of Things (IoT) technology. The data is prone to hack and can make a severe threat to the life of any important persons. This worst situation can be avoided with the help of the Pipelined SMS4-BSK encryption transmission system.

Figure A4 shows the implementation of an automatic Voice Prescription Generation and secured transmission to the patient using Pipelined SMS4-BSK cryptosystem.

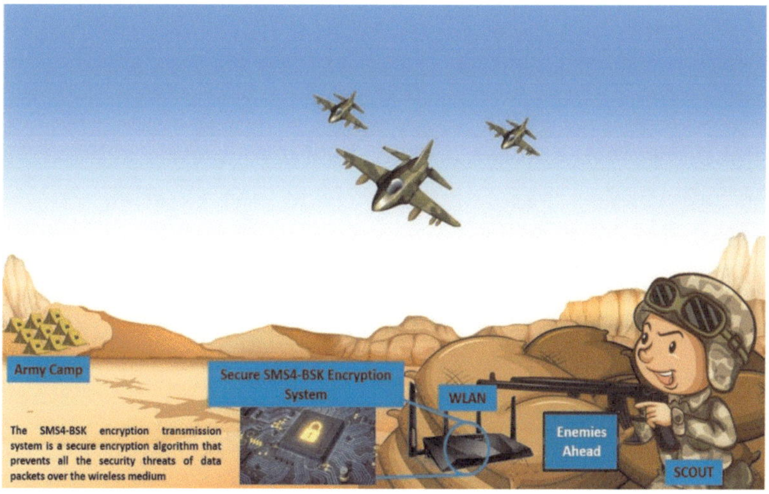

Figure A2.
A war field message transmission.

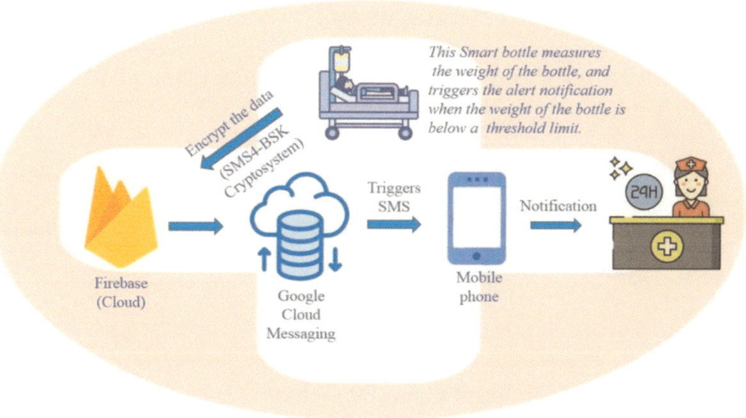

Figure A3.
IoT based smart bottle for healthcare with secure data transmission

Figure A4.
Voice prescription generation with secure communication.

Author details

Babu M.[1*], Sathish Kumar G.A.[2], Gurumurthy J.[1] and Josephine Shermila P.[1]

1 Department of Electronics and Communication Engineering, R.M.K. College of Engineering and Technology, Chennai, India

2 Department of Electronics and Communication Engineering, Sri Venkateswara College of Engineering, Chennai, India

*Address all correspondence to: ibabum@gmail.com

References

[1] Jin Y, Shen H, You R. Implementation of SMS4 Block Cipher on FPGA. In: Proceedings of the First International Conference on Communications and Networking in China; China. 2006. pp. 1-4

[2] Gao X, Lu E, Xian L, Chen H. FPGA Implementation of the SMS4 Block Cipher in the Chinese WAPI Standard. In: Proceedings of the International Conference on Embedded Software and Systems Symposia. 2008. pp. 104-106

[3] Han L, Han J, Zeng X, Ronghua L, Zhao J. A programmable security processor for cryptography algorithms. In: Proceedings of the 9th International Conference on Solid-State and Integrated-Circuit Technology. 2008. pp. 2144-2147

[4] Zhao M, Shou G, Hu Y, Guo Z. High-speed architecture design and implementation for SMS4-GCM. In: Proceedings of the Third International Conference on Communications and Mobile Computing. 2011. pp. 15-18

[5] Zhao J, Guo Z, Zeng X. High throughput implementation of SMS4 on FPGA. IEEE Access. 2019;**7**: 88836-88844. DOI: 10.1109/ ACCESS.2019.2923440

[6] Lee JH, Lee SE, Yu HC, Suh T. Pipelined CPU design with FPGA in teaching computer architecture. IEEE Transactions on Education. 2012;**55**(3): 341-348. DOI: 10.1109/TE.2011.2175227

[7] Abdel-hafeez S, Sawalmeh A, Bataineh S. High performance AES design using pipelining structure over GF((24)2). In: Proceedings of the IEEE International Conference on Signal Processing and Communications. 2007. pp. 716-719. DOI: 10.1109/ICSPC.2007.4728419

[8] Guo Z, Li G, Liu Y. Dynamic reconfigurable implementations of AES algorithm based on pipeline and parallel structure. In: Proceedings of the Second International Conference on Computer and Automation Engineering (ICCAE). 2010. pp. 257-260. DOI: 10.1109/ ICCAE.2010.5451864

[9] Chueng TP, Yusoff ZM, Sha'ameri AZ. Implementation of pipelined data encryption standard (DES) using Altera CPLD. In: TENCON Proceedings. Intelligent Systems and Technologies for the New Millennium. 2000. pp. 17-21. DOI: 10.1109/ TENCON.2000.892211

[10] Babu M, Mukuntharaj C, Saranya S. Pipelined SMS4 Cipher design for fast encryption using twisted BDD S-Box architecture. International Journal of Computer Applications & Information Technology. 2012;**1**(3):26-30

[11] Taherkhani S, Ever E, Gemikonakli O. Implementation of Non-Pipelined and Pipelined Data Encryption Standard (DES) Using Xilinx Virtex-6 FPGA Technology. In: Proceedings of the Tenth IEEE International Conference on Computer and Information Technology. 2010. pp. 1257-1262. DOI: 10.1109/ CIT.2010.227

[12] Patterson DA, Hennessy JL. Computer Organization and Design. 5th ed. USA: Morgan Kaufmann Publishers Inc; 2013. p. 800

[13] Babu M, Sathish Kumar GA. Design of Novel SMS4-BSK encryption transmission system. Integration. 2021;**78**:60-69. DOI: 10.1016/j.vlsi.2021.01.003

[14] Babu M, Sathish Kumar GA. Enhanced moth flame optimization based Supervision Kernel Entropy

Component Analysis for high-speed encrypted transmission model. International Journal of Communication Systems. 2021;**34**(10):1-21

[15] Babu M, Sathish Kumar GA. In depth survey on SMS4 architecture. In: Proceedings of the IEEE International Conference on Intelligent Computing and Communication for Smart World (I2C2SW). Erode, India; 2018. pp. 33-36

Chapter 4

Hardware Implementation of an Improved Hybrid Cryptosystem for Numerical Image Encryption and Authenticity

Amal Hafsa, Jihene Malek and Mohsen Machhout

Abstract

Cryptography is the science that concerns protecting information by transforming its comprehensible form into an incomprehensible one. The conception of a robust cryptosystem is a challenge. In this paper, an improved hybrid cryptosystem for numerical image protection is presented. First, the initial secret key is generated by a secure hash function (keccak). Secondly, the plain image is encrypted through the advanced encryption standard (AES) with CTR mode. Finally, a Rivest-Shamir-Adleman (RSA) algorithm is used to secure the symmetric key transmitted over the insecure channel and owner signature. Our cryptosystem is implemented in hardware and evaluated by different tools mainly identified from the image cryptography community using numerous kinds of standard images. The experimental and analytical findings prove that our framework security gives a trade-off between robustness and performance, which can be used in several domains like medicine, military, and community privacy.

Keywords: hybrid scheme, encryption/decryption, signature/verification

1. Introduction

The digital image is a technology that contains secret information. In today's highly networked world, web servers, application servers, and database backend all connect through public or shared digital networks. In this environment, image is vulnerable to potentially more destructive attacks such as replay or human-based attacks, brute-force, statistical attack, etc. Cryptography is the ideal solution to protect numerical images in storing and moving. It provides a deep defense enough against last record. Evidently, security surrounds us daily at a personal level. Cryptography solutions should guarantee the confidentiality, integrity, and authentication of the secret data. The confidentiality is enabled by encryption that transforms secret data from readable forms to unintelligible forms. Encryption provides a good defense against the new generation of attacks. Even if systems are compromised and the information is taken, encryption can keep it unusable. Integrity is that any change in the transmitted data can be detected at the destination. However, authenticity is that the receiver can verify the identity of the sender. Among encryption schemes, hybrid scheme is considered the ideal idea to protect numerical images that permits both confidentiality and authenticity. Symmetric scheme is effective for large volume data encryption because it is generally hundreds to thousands of execution times faster than asymmetric scheme, but it suffers from secret key distribution. On the other hand, the asymmetric scheme is more secure than the symmetric scheme since it uses different pair keys for encryption than for decryption. However, the hybrid scheme is an innovative idea that combines the symmetric approach for large volume data encryption and the asymmetric approach for secret key exchanging and authentication.

In this paper, we make the following contributions:

- A strong hybrid framework using Keccak, RSA, and AES-256 is suggested, in which all security services are guaranteed.
- Undertake in-depth experimental measurements for several kinds of numerical images with different types, contents, and sizes to evaluate the robustness of the cryptosystem put forward.
- Undertake in-depth evaluation study of the performance of the execution and compare the findings with other works.

The rest of the paper presents the following sections: A survey of existing works is given in Section 2. In Section 3, a preliminary study of Keccak, AES, RSA, and counter encryption mode (CTR) are respectively described. The proposed methodology to create the overall algorithm is detailed in Section 4. Section 5 presents the evaluation and the security analysis of the technique put forward. Finally, the last section gives conclusions and related works.

2. Related works

In this section, we recall diverse related works that studied the encryption algorithms designed for securing digital images.

An image cryptosystem that combines chaos sequences and a modified AES algorithm is put forward in [1]. Firstly, the key is generated by Arnold's chaos sequence. Secondly, the plain image is ciphered by the modified AES and by implementing the round keys produced by the chaos system. In [2], the authors proposed a dynamic AES for numerical image encryption based on the logistic chaotic map and the advanced encryption standard (AES) with Galois mode. Utilizing the logistic mapping, the key is generated and mixed at different stages with the plain data. In [3], the authors put forward a secure framework using a chaos system-based S-box. However, a chaos system is employed to generate three sequences of random numbers, then, an S-box is performed. The image is ciphered by the XOR operation and the sub-byte function.

The main issue of these works is that they suffer from the secure transmission of the symmetric key. In [4], a secure method for color image protection is suggested. It is based on the elliptic curve and AES. Random numbers are generated by the elliptic curve, hence these numbers are employed for generating three maskers to cipher the three components-red, blue, and green-of the image. The drawback of this paper is that all operations are performed sequentially, which can degrade the system. In [5], the authors present a novel idea to protect images against attacks. The initial level is performed by applying the conformal mapping to the private data. Then the image result is encrypted and decrypted using the (RSA) algorithm. Thirty they use less significant bit (LSB) as the hiding method to hide the message inside the cover image. Finally, they propose to compress the image using GZIP. However, the use of an asymmetric algorithm to encrypt big data can degrade the system in terms of execution time. In [6], the authors propose a quantum logistic image cryptosystem that combines both RSA and SHA-3 algorithms. Firstly, the RSA is employed to randomly generate key pairs with private keys and public keys. A fixed matrix is then generated for the confusion. Secondly, the preprocessed image is calculated by the hash function SHA-3 to obtain the clear message that is then stored safely. Using the RSA algorithm, the encrypted message can be performed corresponding to the original message. After mixing both the original and the encrypted messages, the initial conditions of the quantum logistic map are computed using a novel mathematical technique. The main drawback of this paper is the use of an asymmetric algorithm that is a hard algorithm to encrypt the big data.

Our motivation for this work is to guarantee all services of cryptography with low computational time.

3. Preliminary study

The proposed security framework uses the SHA-3 for the initial key generation, the AES algorithm for the whole image encryption–decryption, and the RSA algorithm for the key exchange and authentication. More information on the suggested algorithms is given in this section.

3.1 Keccak

Hash functions get a finite arbitrary length of data as an input argument and produce an output data digest of a fixed length of bits. The keccak hash function is selected by the NIST as the contest winner SHA-3 [7]. The keccak enables the integrity of information such that a tiny alteration in the input data, with one bit, will cause a greatly significant change in the output. As a sequence, each data has its own data digest. It is considered a sponge function because it is based on construction sponge as shown in **Figure 1**. It can produce pseudo-random output with the desired length from an arbitrary length entry.

The keccak algorithm is the permutation f, which is repeatedly applied to a fixed data length ($b = r + c$ bits), where c and r are, respectively, the capacity and the throughput binary. Higher values of c correspond to a higher level of security then those higher values of r improve the speed. The sponge function consists of three phases: the first is the padding phase, where the entry is completed by a padding rule that produces an output length multiple of r, the bit rate. Second, the message is divided into blocks with each of r bits in length such that each block is XOR with the bit rate r. The result will then be concatenated with the capacity c to constitute a state that is an entry of the absorbing phase. In this phase, the state is absorbed by a function f, then, a calculation is repeated for a number determined of rounds. Finally, the output of the absorbent phase is extracted from its state (phase of compression) and constitutes the digest with the desired length. More details are given in [8].

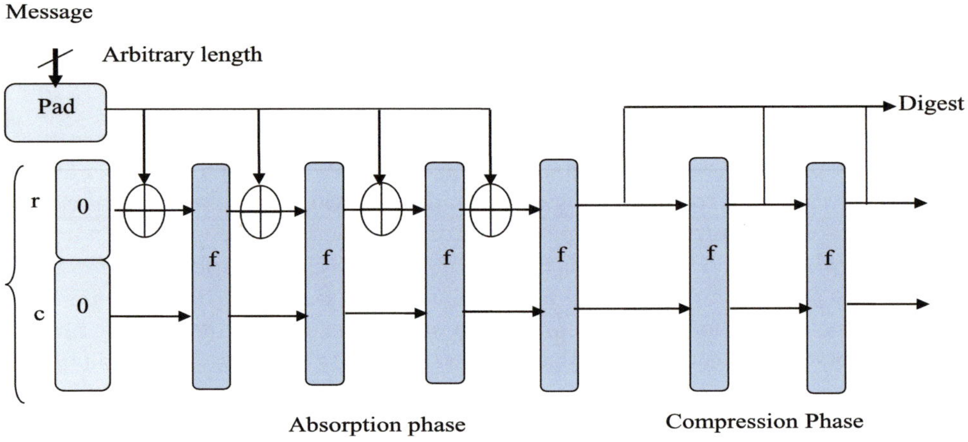

Figure 1.
The construction mechanism of the sponge.

3.2 AES

AES is a symmetric encryption standard considered by the FIPS as the rapid and the more efficient among symmetric algorithms [9]. AES is a block cipher algorithm; data is processed in 128-bit blocks for plaintext and ciphertext. The secret key is 128 bits long, hence the version name is AES 128 (there are two other variants whose keys are 192 and 256 bits). The key with 256 bits of length is the most secure to perform high security. The AES is composed of four main operations to perform both confusion and diffusion for the Shannon Principe, which are the AddRound key, the Subbytes, the Shiftrows, and the MixColumns. The key expansion is used to generate keys from an initial secret key. More information is given in [9]. The flow design of the AES is shown in the **Figure 2**.

3.3 CTR encryption mode

In this mode, the key flow is obtained by encrypting the successive values of a counter. This mode combines many advantages because it allows stream encryption and is pre-computable. In addition, it allows random access to data, is parallelizable, and only uses the encryption function. The counter used can be a pseudo-random sequence that will be easy to find from the seed (initialization vector). Encryption architecture is detailed in **Figure 3**.

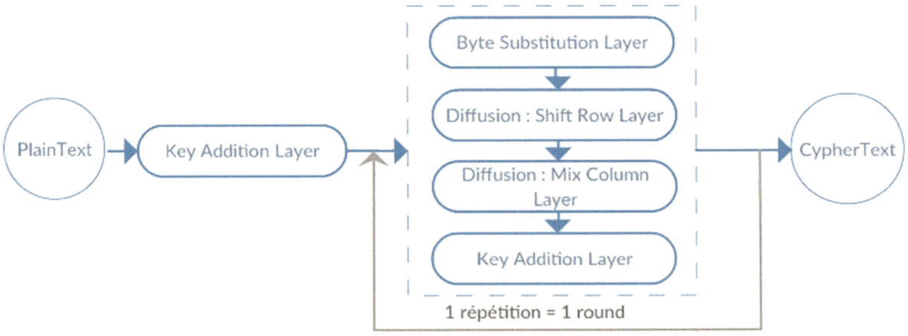

Figure 2.
Flow design of the AES.

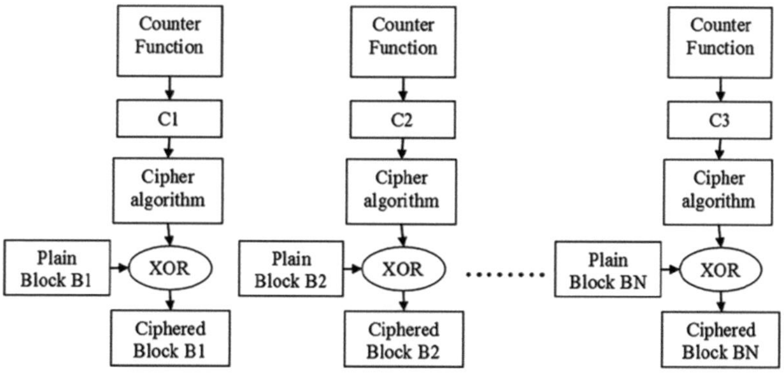

Figure 3.
General architecture of CTR encryption mode.

3.4 RSA algorithm

The RSA cryptosystem provides a digital signature scheme (sign + verify), based on the math of the modular exponentiations and discrete logarithms and the computational difficulty of the RSA problem (and its related integer factorization problem). The RSA sign/verify algorithm works as described below.

1. Key Generation Procedure

    1. Select two distinct big random prime numbers p and q.

    2. Calculate n by multiplying p and q.

    3. Calculate φ(n) = (p-1) (q-1).

    4. Choose e, 1 < e < φ(n), which is relatively prime to φ(n).

    5. Calculate d to perform d ≡ e-1 mod φ(n); d is considered as a private key exponent.

    6. The public key is (n, e) and the private key is (n, d). Keep all the values p, q, and φ(n) private.

2. RSA Sign

    1. Signing a data A with the private key exponent d

    2. Compute the digest of the message: h = hash(A)

    3. Encrypt h to calculate the signature

3. RSA Verify Signature

    1. Verifying a signature ‘s’ for the message A with the public key exponent e

2. Compute the hash: h = hash(A)

3. Decrypt the signature: $h'=se(\mathrm{mod}n)$

4. Compare h with h' to find whether the signature is valid or not

5. If the signature is correct, then the following will be correct:

$$h' = se(\mathrm{mod}\ n) = (\mathrm{hd})\ e(\mathrm{mod}\ n) = h$$

4. Proposed cryptosystem

Hybrid cryptography framework presents the perfect idea for numerical image protection. It employs both symmetric and asymmetric algorithms. The first one is employed for large data encryption and the second one is to share the secret key and permits authentication by signature generation, fully related to the data and the sender. The keccak hash function is used to generate the initial secret key provided as the digest of the original image.

The flow design of the hybrid cryptosystem is given in **Figures 4** and 5.

4.1 Encryption process

The encryption process requires a symmetric algorithm to encrypt the plain image. For this, we have chosen to use the AES-256 algorithm with CTR mode for the encryption process because it is the rapid mode. However, AES requires an initial key Ki for key expansion, so we have proposed to generate the keccak hash function. On the other hand, the image is decomposed into blocks of 32 bytes. As a consequence,

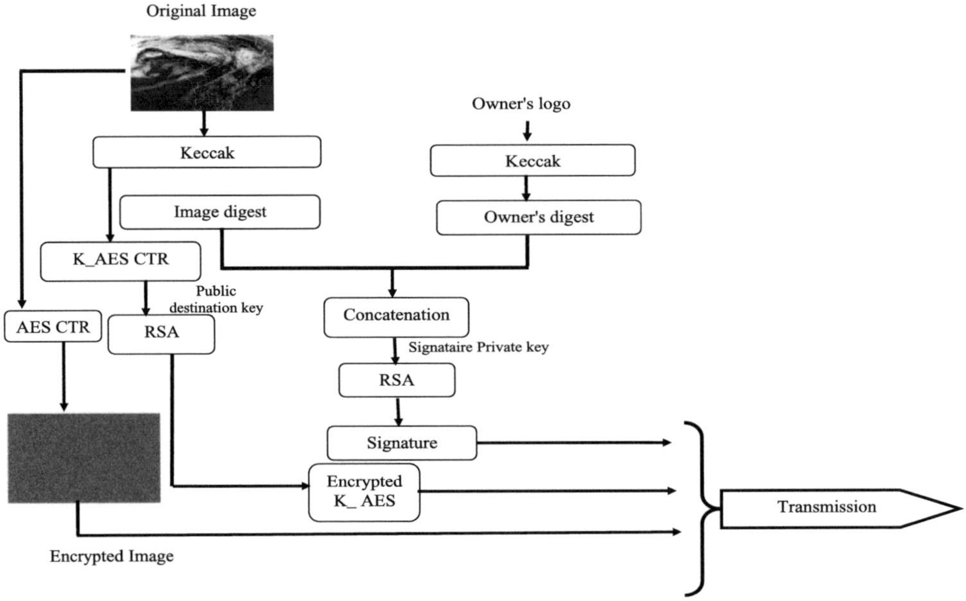

Figure 4.
Encryption and generation of the digital signature.

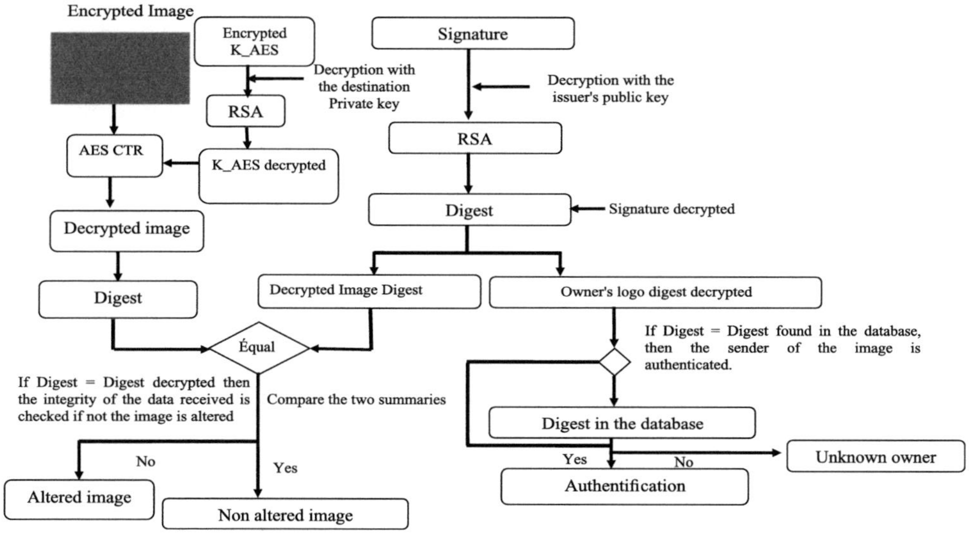

Figure 5.
Decryption and verification of the digital signature.

the used encryption mode is the CTR. Thus, a counter function is employed for generating 256-bit count values, after that, it is encrypted by the AES-256 mechanism. On the other part, to secure the transmission of the initial key Ki, which is ciphered using the RSA algorithm.

4.2 Image authentication and the Key exchange

For image authentication, a 256-bit signature is generated fully related to the owner logo using keccak algorithm. The signature is combined with the 256-bit initial key and encrypted together by the RSA for more security. However, the initial key is encrypted using the public key of the receiver using the Eq. (1). At the reception, the private key is only used for decrypting the initial key, using Eq. (2),

$$K_f = K_i^e \ mod \ (n) \tag{1}$$

where K_i is the plain key, Kf is the correspondent encrypted key, and (n, e) is the public key of destination.

$$K_i = K_f^d \ mod \ (n) \tag{2}$$

where K_i is the decrypted key, K_f is the encrypted key, and (n, d) is the private key of the destination.

4.3 Decryption process

After the encryption process, ciphered image is sent to the destination over an insecure channel. At the reception, a decryption process should be done to restrain the plain image and detect the owner's logo. This phase is the reverse of the encryption process. The algorithm starts by decrypting the encrypted key by the RSA system to obtain the plain key. The first part, 256-bit, is the initial key Ki used for image decryption, and the remaining part, 256-bit, is the owner's digital signature used for image authentication. However, using Ki, the received image is ciphered by the decryption system. This treatment enables finding the original image. On the other hand, using the digital signature with the help of a database, we detect the image owner's logo.

5. Hardware implementation of the proposed hybrid cryptosystem

The suggested hardware architecture is performed with reduced resources that the embedded system makes available. Al cryptographic functions are required to design the hybrid secure framework operated in 32 bits datapath. In our design, six essential blocks are performed with the Control Unit, AES block, the RSA block, keccak block, and I/O buffer for 32-bit data bus. The control unit controls all algorithms and the information exchanged from external devices. Buffer in and buffer out are necessary for communicating the data from and to the on-chip bus. The AES block is used for data encryption and decryption. The keccak block is designed for hashing the message and finally, the RSA block is performed for the signature generation and verification. The proposed hardware architecture is given in **Figure 6**.

The suggested hybrid cryptographic framework is implemented on the DE2–115 board featuring Cyclone IV.E FPGA. **Table 1** gives the utilization of resources when synthesized by the Quartus II tools. The system necessitates 80% of logic elements, 79% of combinational functions, 27% of logic registers, and 7% of memory and consumes 226.15 mW. Concluding the obtained results, the proposed cryptosystem hardware design occupies a small hardware area and consumes reduced power. Thus, the proposed cryptosystem meets the constraints of onboard systems.

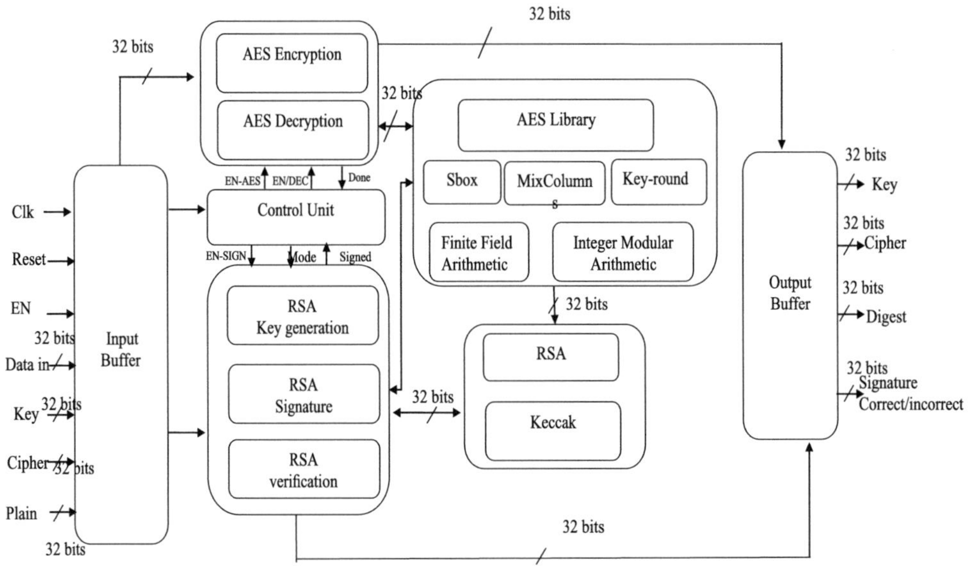

Figure 6.
Hardware architecture of the proposed cryptosystem.

Hardware performances	Resources utilizations
Total logic elements	91.584 (80%)
Total combinational functions	90.439 (79%)
logic registers	30.90 (27%)
Memories	273.408/3981312 (7%)
Total power consumption (mW)	226.15

Table 1.
Hardware resources results.

6. Evaluation and security analysis

To prove the robustness of the proposed security framework, we evaluate it using various metrics or tools for different standard images. This section contains the statistical analysis, the key analysis, Know Plain Text and Chosen Plain Text Attack, and the speed.

6.1 Statistical analysis

Statistical analysis indicates the factor of coincidence between plain and cipher images to test the ability of the algorithm to defend against attacks. Statistical analysis can include the image histogram, the entropy, the normalized correlation, the correlation coefficient (ρ), the peak signal to noise ratio (*PSNR*), and structural similarity index measure (*SSIM*) tools.

6.1.1 Histogram analysis

The histogram represents the data distribution in function of the pixel's values. A perfect secure algorithm must create a cipher image with uniform and totally different histograms compared to the original images. We analyze the histograms of original and encrypted samples and the contents are very different as shown in **Figure 7**. As a consequence, it is so hard to understand the encrypted image's appearance.

6.1.2 PSNR, SSIM, and entropy analysis

After encryption, the entropy factor, the PSNR, and SSIM are calculated as reported in **Table 1**. Practically, a perfect cryptosystem performs random information equal to "8" [10]. As given in **Table 2**, we constate that the entropy value of the cipher sample is close to the ideal value "8." This means that our algorithm put forward can resist statistical attacks. When analyzing the PSNR results. We note that the value is lower than 5 dB (<5 dB). Following the reference [11], the suggested cryptosystem gives a bad quality between both plain and cipher samples. Thus, it is hard to predict the original image from the cipher one. When turning to the SSIM factors computed between the original and the cipher images, we constate that values are close to 0. Therefore, we cannot extract the content of the clear sample from the cipher one. A comparative study of other existing work for entropy and PSNR evaluation tools is

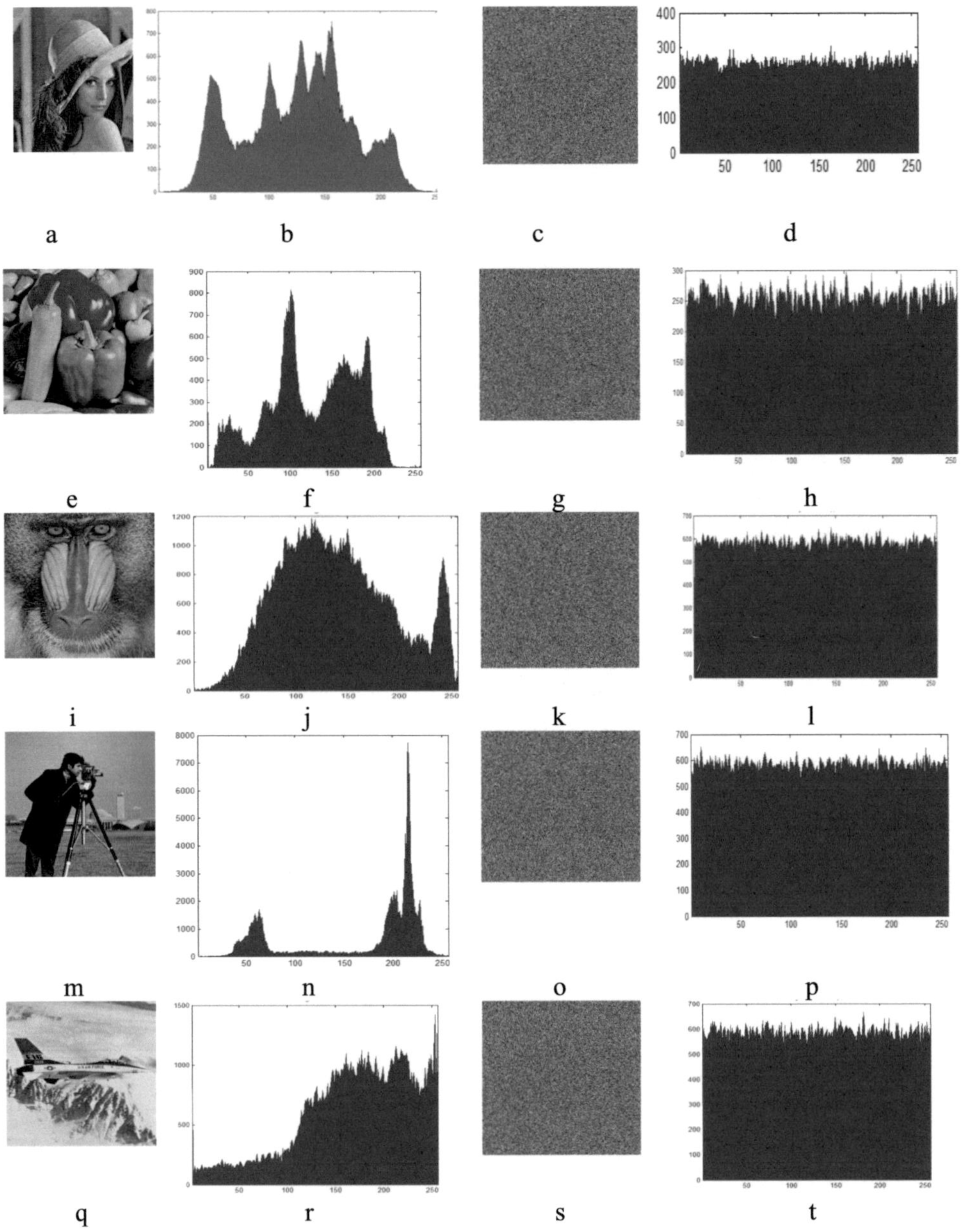

Figure 7.
Histogram of the plain samples and their corresponding of the cipher samples.

reported in **Table 3**. This comparison lets us conclude that our cryptographic model outperforms the other works and can resist statistical attacks.

6.1.3 Correlation coefficient analysis

The standard plain sample is characterized by a correlation close to 1. Although in a cipher sample the adjacent pixels should be uncorrelated [15]. Let x and y two gray

Standard image	E	PSNR	SSIM
Lena	7.99975	4.521	0.0101
Peppers	7.99972	4.765	0.0089
Baboon	7.99972	4.580	0.0102
Walkbridge	7.99975	4.769	0.0086
Cameraman	7.99972	4.395	0.0089
Jetplane	7.99936	4.378	0.0105

Table 2.
E, PSNR, SSIM, and NC values of cipher samples.

Algorithm	PSNR	E
Ref. [12]	_	7.99920
Ref. [3]	_	7.99122
Ref. [13]	10.0454	7.75970
Ref. [6]	–	7.9993
Ref. [14]	0.9756	7.5631
Proposed algorithm	4.521	7.99975

Table 3.
Comparative study of PSNR and E values for Lena image.

scale parameters of two adjacent pixels in the sample, and the correlation of the adjacent pixels is evaluated by the following Eqs. (3)–(6).

$$E(x) = \frac{1}{N}\sum_{i=1}^{N} x_i \tag{3}$$

$$D(x) = \frac{1}{N}\sum_{i=1}^{N} (x_i - E(x_i))^2 \tag{4}$$

$$cov(x,y) = \frac{1}{N}\sum_{i=1}^{N} (x_i - E(x))(y_i - E(y)) \tag{5}$$

$$r_{xy} = \frac{cov(x,y)}{\sqrt{D(x)}\sqrt{D(y)}} \tag{6}$$

Table 4 demonstrates that the cipher sample's correlation coefficients are close to 0. Therefore, we cannot extract the content of the plain sample from the cipher one. **Figure 8** illustrates the distributions of 2000 pairs of randomly selected adjacent pixels of the clear and ciphered Lena image.

Findings show that the correlation coefficient of the clear samples is close to 1, while the cipher images are close to zeros. Similarly, the distribution of adjacent pixels is inconsistent; i.e., there is no correlation between them. This proves that the cryptosystem eliminates the correlation of adjacent pixels in the clear sample, and it makes a ciphered sample with no correlation. A comparative evaluation with some existing

Image	Status	Horizontal	Vertical	Diagonal
Lena	Original	0.80482	0.81942	0.82776
	Cipher	0.000359	−0.0001339	−0.007713
Peppers	Original	0.84679	0.97335	0.88347
	Cipher	0.000638	−0.06850	−0.05433
Baboon	Original	0.83045	0.87679	0.87929
	Cipher	0.000757	−0.000602	−0.0010326
Walkbridge	Original	0.87108	0.87934	0.84997
	Cipher	0.00093	0.00380	−0.00776
Cameraman	Original	0.87790	0.89079	0.90087
	Cipher	0.000216	−0.00774	−0.00580
Jetplane	Original	0.80839	0.81643	0.82083
	Cipher	−0.005928	0.00381	−0.0015028

Table 4.
ρ values of clear image and its correspondent cipher image.

work for the correlation coefficient is tabulated in **Table 5**. The comparison proves that our system put forward gives the perfect results. Thus, the system can resist the statistical attacks.

6.2 Key analysis

To validate the robustness of the suggested algorithm, the key space, the key sensitivity, and the randomness analysis are tested in this section.

6.2.1 Key space

According to [18–20], the key space of a robust cryptosystem must be large to be protected from the brute-force hacker. In our algorithm, for an initial key Ki, there are 2^{128} dissimilar keys, which are very large. Therefore, the key brute-force attacks are computationally infeasible.

6.2.2 Key sensitivity

To assure a high level of protection, the cryptographic algorithm must be sensitive to the sample input and the initial secret key Ki. A simple alteration (one bit) in Ki, or in the sample, will cause a greatly significant modification in the output generated keys for encryption and output sample. The parameter Key sensitivity can be performed by employing the number of changing pixel rate (*NPCR*) and unified averaged changed intensity (*UACI*) randomness tests to test the force of the algorithm to defend against differential attacks [20], which are described as follows:

$$NPCR = \frac{1}{S}\sum D(i,j) \times 100\% \quad (7)$$

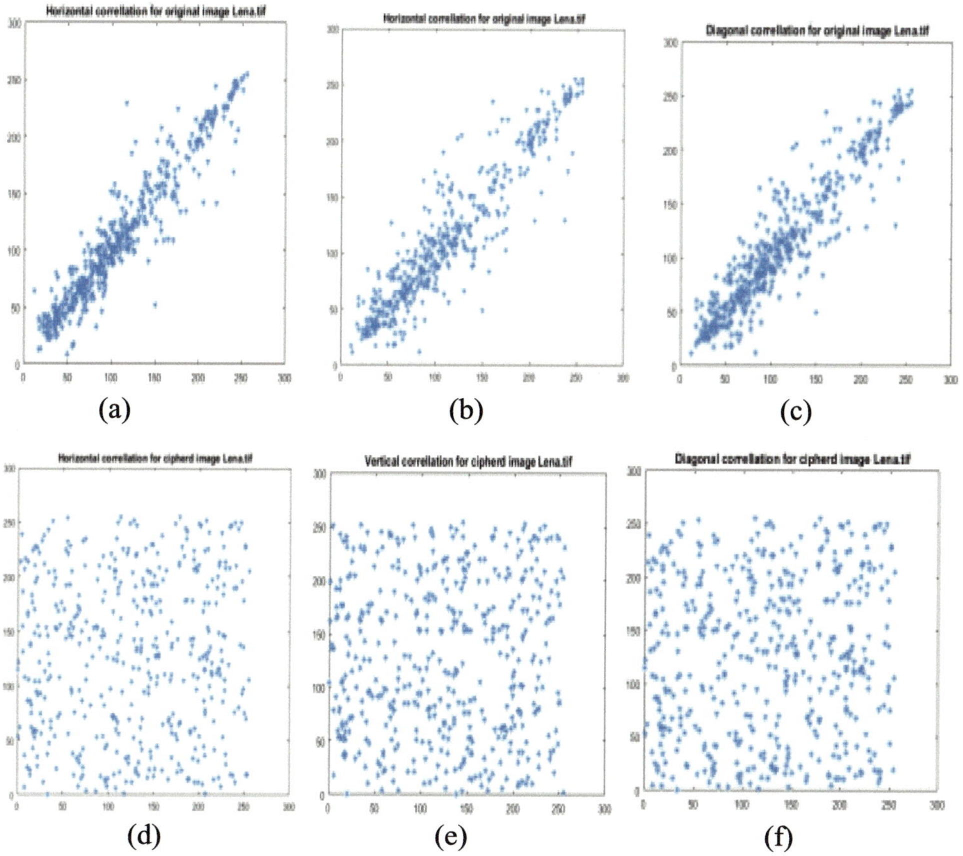

Figure 8.
Distribution of 2000 pairs of randomly chosen adjacent pixels for Lena image: (a)–(c): horizontal, vertical, and diagonal distribution of the original image; (d)–(f): horizontal, vertical, and diagonal distribution of the cipher image.

Work	Horizontal	Vertical	Diagonal
Ref. [12]	0.0265	0.0792	0.0625
Ref. [3]	−0.0001	0.0089	0.0091
Ref. [15]	0.0036	0.0035	0.0064
Ref. [16]	0.0040	0.0015	0.030
Ref. [17]	0.0002	−0.0133	−0.0791
Proposed system	0.0003	−0.0001	−0.00771

Table 5.
Comparative study of correlation coefficient for Lena image.

$$UACI = \frac{1}{S}\sum \frac{|d|}{G} \times 100\% \tag{8}$$

where S is the size of the image, and $D(i, j)$ is a logical value affected by the following cases:

$$D(i,j) = \begin{cases} 0 \; if \; I_1(i,j) = I_2(i,j) \\ 1 \; if \; I_1(i,j) \neq I_2(i,j) \end{cases} \quad (9)$$

d is the variance between two pixels on the sample with the same coordinates.

$$d = p_1(i,j) - p_2(i,j) \quad (10)$$

Thus, a sensitivity test applied to the initial key is evaluated by two initial keys, Ki1 and Ki2, where the key Ki2 is dissimilar by only one bit from the initial key Ki1 to cipher the same input. Then, we try deciphering the obtained samples with a wrong key. Here, the two keys, Ki1 and Ki2, are permuted in the decryption step; i.e., each image is decrypted by a wrong key, which is different by one bit from the correct key. This test is carried out with many Ki keys, which are randomly selected to properly evaluate the algorithm. Simulation results for the Lena image are shown in **Figure 9** and **Table 6**.

When analyzing results, we can conclude that our encryption system is very sensitive to small modifications in the entered sample. This proves the efficacity of the keccak hash function, which puts an image's initial key specific for encryption.

A comparative study is given in **Table 7** to evaluate the system compared to other related work and the results prove that the system is robust.

6.3 Know plain text and chosen plain text attack

This type of hacker is used to crack some of the cryptographic algorithms. Usually, an adversary employs black or white samples to extract the possible patterns in the algorithm. The white and dark samples are encrypted by the proposed method.

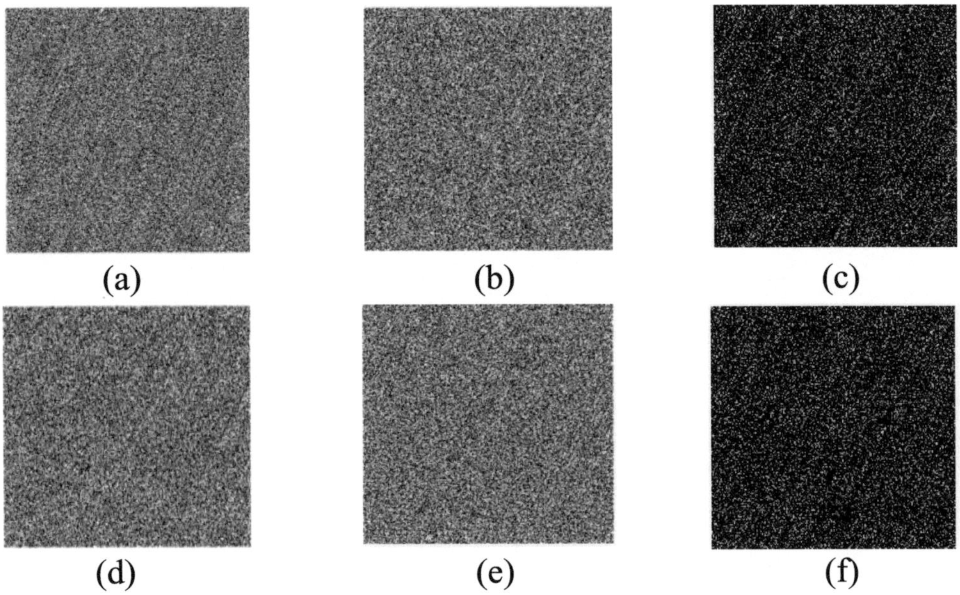

Figure 9.
Key sensitivity test applied on initial key, (a) ciphered Lena by Ki_1, (b) ciphered image by key Ki_2, (c) variance between (a) and (b), (d) deciphered (a) by Ki_2, (e) deciphered (b) by Ki_1, and (f) variance between (d) and (e).

Image	*NPCR* (%)	*UACI* (%)
Lena	99.7129	33.2936
Peppers	99.7698	33.2384
Baboon	99.75895	33.9487
Walkbridge	99.6789	33.45796
Cameraman	99.69875	33.60235
Jetplane	99.76589	33.38546

Table 6.
NPCR *and* UACI *tests applied on the initial key.*

Methods	*NPCR* (%)	*UACI* (%)
Ref. [21]	99.63230	33.12500
Ref. [22]	99.61000	33.53000
Ref. [23]	100.0000	33.43150
Ref. [24]	99.58948	33.46458
Ref. [25]	99.62530	33.48070
Proposed algorithm	99.7129	33.2936

Table 7.
Comparative evaluation of NPCR and UACI parameters for Lena image.

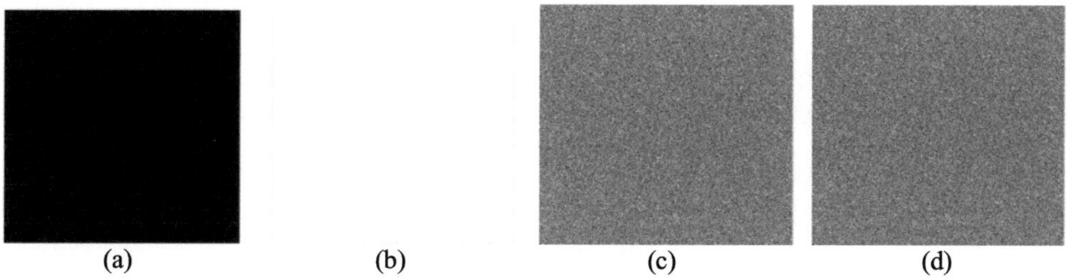

Figure 10.
(a) Black image, (b) white image, (c) ciphered black image, (d) ciphered white image.

Figure 10 gives the encrypted samples and no pattern is apparent. The value of the entropy for samples is selfsame as in other images and correlation coefficients are perfect. **Table 8** tabulates the correlation between adjacent pixels and the entropy values of both samples. Our proposed hybrid scheme is very resistant to attacks.

6.4 Encryption algorithm speed

In real-time application, the run time is considered as a main constraint. Eqs. (11) and (12) are used to calculate the speed (S) and the number of cycles per byte obtained by a specific algorithm running on the processor.

Image	Correlation coefficients			
	Entropy	Horizontal	Vertical	Diagonal
Black	0	—	—	—
Cipher Black	7.99975	−0.0026	0.00079	0.0002
White	0	—	—	—
Cipher White	7.99975	0.00026	−0.00201	0.00056

Table 8.
Entropy and correlation values.

Work	Used tools	*S (MB/S)*	*CpB*
Ref. [4]	Elliptic curve + AES CBC	0.002	1,400,000
Ref. [26]	Logistic-Tent and Tent-Sine	0.19	10,520
Ref. [27]	Chaos system + DWT	0.019	147,368
Ref. [28]	SHA2 + AES + RSA	0.38	8947
Proposed Algorithm	KECCAK +AES-CTR+ RSA	0.45	7.555

Table 9.
Performance results and comparison with the state of the art.

$$S = \frac{data\ size}{Time} MB/s \tag{11}$$

$$number\ of\ cycles/byte = \frac{Frequency}{S} \tag{12}$$

A comparison with related works is given in this state (**Table 9**). The comparison proves that the suggested scheme has the best findings in terms of speed.

7. Conclusion and future work

In this paper, we have put forward a strong hybrid cryptographic framework for image encryption, decryption, and authentication. The algorithm has combined a hash function and symmetric and asymmetric algorithms. The keccak hash function has been used for the initial secret key generation related to the plain image and owner's signature. The RSA encryption system has been used for the secret key exchanging and authentication. For image encryption, we have utilized the AES-256 bits with CTR mode. The main advantages of this mode are the rapidity of treatment and the no error propagation. The evaluation and analysis results prove that our proposed algorithm allows high performance and security. It can resist most known cryptanalysis attacks. For future work, we aim to design a mechanism for dynamically changing the S-box values of the AES.

Author details

Amal Hafsa[1]*, Jihene Malek[1,2] and Mohsen Machhout[1]

1 Electronics and Micro-Electronics Laboratory, University of Monastir, Monastir, Tunisia

2 Department of Electronics, Sousse University, Higher Institute of Applied Sciences and Technology, Sousse, Tunisia

*Address all correspondence to: hafsaamal12@gmail.com

References

[1] Arab A, Rostami MJ, Ghavami B. An image encryption method based on chaos system and AES algorithm. The Journal of Supercomputing. 2019;**75**: 6663-6682. DOI: 10.1007/s11227-019-02878-7

[2] Shariat M, Mohammad Z, Rostami J, Eftekhari M. Proposing a novel dynamic AES for image encryption using a chaotic map key management approach. Optik. 2021;**246**:167779. DOI: 10.1016/j.ijleo.2021

[3] Unal C. Secure image encryption algorithm design using a novel chaos-based S-Box. Chaos, Solitons Fractals. 2017;**95**:92-101

[4] Toughi S. An image encryption scheme based on elliptic curve pseudo random and advanced encryption system. SIG Process. 2017;**141**:217-227

[5] Alawsi WA, Obayes HK, Hussain SM. A novel image encryption approach for IoT applications. Webology. 2022;**19**. DOI: 10.14704/WEB/V19I1/WEB19107

[6] Ye G, Jiao K, Huang X. Quantum logistic image encryption algorithm based on SHA-3 and RSA. Nonlinear Dynamics. 2021;**104**:2807-2827. DOI: 10.1007/s11071-021-06422-2

[7] Sideris A, Sanida T, Dasygenis M. High throughput pipelined implementation of the SHA-3 cryptoprocessor. In: 2020 32nd International Conference on Microelectronics (ICM); Aqaba, Jordan. 2020. pp. 1-4. DOI: 10.1109/ICM50269.2020.9331803

[8] Christof Paar, Jan Pelzl, SHA-3 and the Hash Function Keccak, An Extension Chapter for Understanding Cryptography—A Textbook for Students and Practitioners Springer, (2012). Available from: www.crypto-textbook.comm

[9] FIPS PUB 197: Advanced Encryption Standard (AES). Computer Security Standard, Cryptography. 2001

[10] Dridi M et al. Cryptography of medical images based on a combination between chaotic and neural network. Journal of Image Processing IET. 2016; **11**(5):324-332

[11] Melo A et al. Chapter 11: PriorityQoE: Atool for Improving the QoE in Video Streaming. Intelligent Multimedia Technologies for Networking Applications: Techniques and Tools. 2013. DOI: 10.4018/978-1-4666-2833-5

[12] Xiuli C et al. A color image cryptosystem based on dynamic DNA encryption and chaos. Journal of Signal Processing. 2019;**155**:44-62. DOI: 10.1016/j.sigpro.2018.09.029

[13] Suri S et al. An AES–CHAOS-Based Hybrid Approach to Encrypt Multiple Images, International Conference in Recent Developments in Intelligent Computing, Communication and Devices. Springer; 2017. pp. 37-43. DOI: 10.1007/978-981-10-3779-5_6

[14] Shi M, Guo S, Song X, Zhou Y, Wang E. Visual secure image encryption scheme based on compressed sensing and regional energy. Entropy (Basel). 2021;**13**. DOI: 10.3390/e23050570

[15] Zhang W, Wong K -w, Yu H, Zhu Z-l. An image encryption scheme using reverse 2-dimentional chaotic map and dependent diffusion. Communications in Nonlinear Science and Numerical Simulation. 2013;**18**:2066-2080. DOI: 10.1016/j.cnsns.2012.12.012

[16] Kaur M, Singh D. Multiobjective evolutionary optimization techniques based hyperchaotic map and their applications in image encryption. Multidimensional Systems and Signal Processing. 2021;**32**:281-301. DOI: 10.1007/s11045-020-00739-8

[17] Gafsi M, Hajjaji MA, Malek J, Mtibaa A. Efficient encryption system for numerical image safe transmission. Journal of Electrical and Computer Engineering. 2020;**8937676**:12. DOI: 10.1155/2020/8937676

[18] Hafsa A, Sghaier A, Malek J, Machhout M. Image encryption method based on improved ECC and modified AES algorithm. Multimedia Tools and Applications. 2021;**80**:19769-19801. DOI: 10.1007/s11042-021-10700-x

[19] Hafsa A, Gafsi M, Malek J, Machhout M. FPGA implementation of improved security approach for medical image encryption and decryption. Scientific Programming. 2021;**2021**: Article ID 6610655, 20 p. DOI: 10.1155/2021/6610655

[20] Hafsa A, Fradi M, Sghaier A, et al. Real-time video security system using chaos-improved advanced encryption standard (IAES). Multimedia Tools and Applications. 2022;**81**:2275-2298. DOI: 10.1007/s11042-021-11668-4

[21] Ramzi G et al. A novel chaos-based image encryption using DNA sequence operation and Secure Hash Algorithm SHA-2. Nonlinear Dynamics. 2016;**83**(3): 1123-1136

[22] Jiahui W et al. Color image encryption based on chaotic systems and elliptic curve ElGamal scheme. Journal of Signal Processing. 2017;**141**:109-124

[23] Khalaf A. Fast image encryption based on random image key. International Journal of Computer Applications. 2016;**3**:0975-8887

[24] Akram B et al. Chaos-based partial image encryption scheme based on linear fractional and lifting wavelet transforms. Journal of Optics and Lasers in Engineering. 2017;**88**:37-50

[25] Rim Z et al. Image encryption based on new beta chaotic maps. Journal of Optics and Lasers in Engineering Elsevier. 2017;**96**:39-49

[26] Ye H-S et al. Multi-image compression-encryption scheme based on quaternion discrete fractional Hartley transform and improved pixel adaptive diffusion. Journal of Signal Processing. 2020;**175**. DOI: 10.1016/j.sigpro.2020.107652

[27] Ahmed N. Timing filter for counter mode encryption. In: 2013 2nd National Conference on Information Assurance (NCIA). 2013. DOI: 10.1109/ncia.2013.6725333

[28] Gafsi M, Malek J, Ajili S, Hajjaji MA, Mtibaa A. High securing cryptography system for digital image transmission. SETIT 2018 SIST. 2020;**146**:311-322. DOI: 10.1007/978-3-030-21005-2_30

Section 2

Cybersecurity Approaches and Keyless Encryption

Chapter 5

Intelligent Cybersecurity Threat Management in Modern Information Technologies Systems

Mohammed Saeed Jawad and Mohammed Hlayel

Abstract

Recently, cybersecurity threat management policy is important to be integrated in the management of any organization using information systems whether these organizations are big, medium or even small. As a good practice to be adopted, these organizations need to adapt to recent trends of security threats to prevent these threat or to minimize the risks associated to them. Understanding attackers' behavior in crucial in the success of this journey and it's always good to probe the systems as ethical hacking to identify possible security vulnerabilities and points of attacks. Modern information systems as cloud computing even should be considered in special care based on the characteristics of cloud security as data confidentiality, encryption and availability in the context of agile DevOps software project management. This book chapter presents the best practices to be adopted and how the organization adapts by setting realistic and reasonable security policies to intelligently manage different types of security threats.

Keywords: cybersecurity, intelligent threat management, security vulnerabilities, network traffic, service ports, hackers, malicious attacks, security defense

1. Introduction

Cybersecurity intelligent threat analysis and management is crucial in today modern information systems. Simply that can be explained as security attacks and threats are becoming more complicated, sometimes difficult to be detected and highly diverse in their characteristics and types. Cybersecurity analysts should always upgrade their skills and knowledge to cope with latest advances in threat attacks. They need to well-understand the different types of system vulnerabilities that attackers may use to hack the victim systems. Therefore, it is always important to have updated databases about different threat techniques, characteristics and their intended objectives. Hackers are usually patient to use different tools to understand and navigate through certain distributed systems that are under analysis for intended attacks based on the discovered system vulnerabilities. It is always good practices to understand the processes of hacking and the tools can be used to analyze the system and find the security weaknesses. As recommendations to leverage the security levels and maturity in organization, it is always important to real-time monitoring of the different system's open services and automatically detect any unusual traffic patterns that may indicate possible attacks on the system. It brief, this situation can be described as continuous racing between attackers and cybersecurity analyst teams to lunch possible attacks from one side and to close these attacks or at least stop spreading them and minimize their risks also recover the system from these malicious harmful activities.

This chapter navigate through recent threat attacks based on definitions, characteristics, tools used to make them feasible and how different types of distributed computing system can be analyzed to scan possible system vulnerabilities. This chapter also discusses the effects of successful attacks and how to minimize these attacks' risks to not allow at the end the malicious attackers achieve their objectives. The aims of this chapter are to have clear knowledge on modern security threats and then to convert that to recommendations and policies in organizations for highly security defense strategies and risk managements to response directly to possible attacks and save the organization valuable data and infrastructure assets.

Cyber security aims to protect the information and systems which consider the valuable assets of any organization. The three most important aspects of the information security are namely the confidentiality, integrity and availability. Confidentiality means only the authorized individuals have the right to access information and resources [1]. Considerable parts of malicious attacks are involving on disclosure to make sensitive information available to the public. Integrity concerns about making an organization's information away from unauthorized changes and these changes can be usually by hacker intestinally or due to services disruption which happen accidently and resulting in changing the system's stored data. Information security also highly concern about the data availability so that authorized individuals can access required information when needed [2]. There are types of security attacks aiming to undermine availability of data such as denial-of-service attack.

This chapter is organized in six sections to fully investigate the different types of malware threats in Section 2. In Section 3 development cycle as tools management and information threat intelligence in fully explained then, threat modeling as risk management for effective threat hunting is highlighted in Section 4. The different aspects of environmental reconnaissance are fully investigated in Section 5. Finally, cloud-based security services as identity and access management (IAM), data loss prevention, web security, e-mail service security and event management. Therefore, this chapter can be considered as a roadmap for investigating modern security threats and how to intelligently perform detection, prevention and risk mitigation/management of these security threats.

2. Malware threats

In the fields of Information and cyber security; it is highly important to understand the different characteristics of different malicious threats as how they can be propagated in systems and the malicious actions they perform. The discussion here is mainly about viruses, worms and Trojans horses. Computer viruses are named so as they behave similar to the biological virus [3]. They spread from one system to another based on some types of users' actions such as opening an email attachment or even clicking a link in a malicious website also, inserting an infected USB drive in a system. In another way it can be said that viruses cannot be spread unless someone give them the hand and based on that, the best way of protection from viruses is educating other about them.

Second famous malware is the worms and they spread between system without even user interactions. Worms attack through the system vulnerabilities and once any system been affected the worm can use this system to spread to other connected network systems. As worms spread through the system's vulnerabilities, the best way to protect against worms is to keep the systems and application updated to the most recent patches. Worms recently became more aggressive to the level they can cross the virtual barriers and cause serious physical damages.

Trojan horses pretend to be legitimate and beneficial piece of software to attract users to download or install and once the user run the programs then they behave in weird ways. Actually these malicious Trojan horses carry a harmful payload that can do unwanted actions behind the scenes [4]. The best way of protecting from these type of Trojan horses is the application control policy the can be approved by administrators. It is worthy to mention here that; remote access Trojan know by short name as RATs are special types of Trojan horses that allow hackers to remotely control affected systems.

2.1 Adware, spyware and ransomware

Each malware can be characterized by its specific propagation mechanism that determine how it spread to other systems also its payload that determine how it delivers its malicious contents in the infected system [5]. This section investigates the different three types of payloads for Adware, Spyware and Ransomware. Beginning with Adware which a common source method to generate revenues from online. Usually this adverting method is quite legitimate and allowing people to generate income from advertised online content. Unfortunately, this is also an opportunity for malware so, adware can consider a specific purpose malware that displaying advertisement to generate money for the malware author instead of the content owner. Some of the tricky mechanism can be used here by adware as for example directing search queries to search engine that is controlled by the malware author also, displaying pop-up while browsing and some cases changing the legitimate ads from content owner by content benefiting the malware author. This is for sure very harmful for the content author as it destroys their customers' trust.

The second type of famous malware payload is the spyware. This malware generates information without the user's permission and knowledge and send information to the malware author to use in any type of malicious action such as stealing account information or for identity theft [6]. Different techniques can be used as spyware such as keystroke loggers capture to trace user's presses. In some cases, spyware monitor web sites visit to capture the usernames and passwords in accessing some sensitive web sites such as banks. These days monitoring web sites browsing is a common spyware activity that can be used to target advertising to specific users. Most dangerously, some spyware malwares can reach inside the system and are able to scan the hard drives and the cloud storage services to capture sensitive information like some social security numbers that can be useful at the end for identity theft.

The third famous type of malware is the ransomware. This type blocks the legitimate user's access and use of a computer or data until ransom is paid. Encrypting files with secret key to be sold later is the most common way used in this malware. Recently, a very good example of ransomware is the CryptoLocker which is started 2014 and still used until today. Most commonly it arrivers to user's email as an attachment in email message and once the user open that attachment, CryptoLocker starts encrypt files in the hardware using strong RSA encryption algorithms. The encrypted file may include office documents, image and CAD modeling which are considered the most important for the end-user. Usually the malware author has a dedicated control server to keep these encryption keys of the files. Then a deadline will be given to pay a ransom. Recently some surveys show that over 40% of the infected people or organization by CryptoLocker paid the ransom. In this context, it is worthy to mention also another type of malware called scareware which is similar to ransomware but it considered a bluff [6, 7]. It usually pops up messages as a sort of website adverting that designed to warn users about some security issues and to scare the end-users by telling them that their systems are compromised and then offer them solutions. The truth is no security issues and their offered solution is just a fake.

2.2 Malware mechanism as backdoor and logic bombs

The Previously discussed malwares are all have common thing as they are independent program developed and they are developed for malicious actions. Some malwares are not fit into this category as they are independent programs but code chunks that can be inserted into other applications for bad intentions [8–10]. Most common malware in this category are backdoors and logic bombs [8] Backdoors appear when programmer give themselves or others means for future access to a system. The tricks can be used here by simply making programming easier in ways to avoid logging in with user credentials or some mechanisms to allow access later when it happens that users accidently lock themselves out from their systems. End-users usually they not want the vendor to gain access to the system by installing these scripts or piece of code especially that backdoors might be fallen into the wrong persons' hands. Backdoors can be happened with different ways as hardcoded account as certain user name and password are always used to provide access to the system. In some scenarios, when the end-users always use the default password as users forget or not bother to change and finally, there is always possibility to unknown access channels that can access the system by avoiding usual authentication processes. Logic bombs is another kind of malware that works by modifying existing code. In this type, some certain conditions (as specific date or time occurring, specific information in the file's content or specific results from API calls) can be coded to trigger payload of harmful action to be executed [9, 10]. Logic bombs scenarios examples can be explained as certain programmer employee created malicious payload that it is inactivated as long as he/she appears in the system daily, once he/she disappeared as being fired for example, the harmful payload then can be activated.

2.3 Advanced malware techniques

These type of malware are developed to be escaped from being detected from normal anti-malware defense systems. The three good examples here are, rootkits, polymorphism and armored types of viruses. In the first type, the virus is designed to hack the root account which is the super user account which has an unrestricted access to the resources of the system and these privileges are usually preserved to system administrator [11]. After the hacker succeed in gaining access to normal user account then use the rootkit to move to the unrestricted super user access. The concept of this type of viruses can be explained as a technique that uses software techniques hiding other software to hack the system. A variety of malicious payload can be delivered by rootkits as backdoors, botnet agents and adware or spyware. Rootkits can attack the system in both levels of user modes or kernel mode and there is trade-off issue in each of these two modes. The user mode rootkit runs with the normal user privileges and they considered easy to write and difficult to detect however, in the privileges mode, the access to the system is in much advanced privileges with trade-off here that these viruses are difficult to write and relatively are easy to be detected.

The polymorphic viruses are advanced types of viruses that have the ability to fight signature detection. Viruses signature detection is very important concept for isolating viruses by discover their patterns and match that with known code pattern stored in dedicated database. Polymorphic viruses escaping signature detection by changing their behavior from time to time so that the virus files look different in each system been attacked so that no signature matching and that will inactivate the signature detection method. One clear method that polymorphic viruses use is different encryption with different key in each system being attacked to make the virus file look totally different. The virus loader has the decryption key that can retrieve the original virus code.

The third advanced type of malware known as armored viruses came with the ability to stop reverse engineering techniques which are usually used to analyze deeply the viruses at the level of the machine language or the assembly code that considered the DNA of the attacking viruses. The techniques followed in these type of viruses are writing the code in obfuscated assembly language that hide the true intention of the code sometimes also blocking the system debugger and using some techniques to stop the methods of sandboxing that used to isolate viruses.

2.4 Botnet

The concept of botnet can be understood as taking control of computer network using let us say worm viruses propagating through the network as startup from single infected computer. The hacker intention in affecting these network system is to steal the system power, storage or even the network connectivity by joining the infected system with botnets. These botnets can be considered as a collection of zombie computers been connected for malicious actions [12]. Once the hacker succeeds in hacking and gaining control of particular system in any technique discussed earlier then, he/she will join the network to the botnet. The infected network will be considered a victim and will be waiting for further instruction from the hacker. The hacker usually sells or rent this botnet to others to use them for spam delivery or distribution denial-of-services attacks, exposing the system for brute-force attacks to crack passwords or even mine bitcoins activities. It can be said here the key resources of these infected systems as storage, computing power or network connectivity are stolen maliciously. Usually in these type of scenarios the hackers not communicate directly with the infected system to avoid the risk of being discovered by security analyst team that will cut-off these connection of network once been discovered instead, hackers use indirect commands and control mechanisms to hide their true locations and usually the hackers here use punch of these techniques together at the same time to be able to gain control of the botnet to the longest possible time.

3. Threat intelligence

Threat intelligence can be described as the adoption of best practices in identifying latest security threats and the risks associated to these threats that can affect the functionaries and operations in an organization [13]. Threat intelligence is very important part of any organization cyber-security analysis program. Threat intelligence if been applied effectively it can significantly enhance organization security by making this organization updated to the latest security threats and how to deal with these risks fast and accurately. As security threat intelligent analyst, few points should be considered for this type of this highly demanded job recently as the following: First, effective information gathering from trusted common open-source like security web sites, news media, social medias and government-sponsored centers for security analysis as well as security research centers. With such information pinching security attacks can be performed to test the readiness level and the effectiveness of the security defense lines [14].

Threat intelligent centers can play very effective roles to leverage the readiness and maturity levels against latest threats trends by the engagement with security information briefing summarizing latest security critical issues. Examples can be varied as educating audience about IP reputation services to provide real-time information about IP involved on suspected security attacks. This useful information can be sent directly to be utilized in firewalls, intrusion prevention and other security tools. Threat intelligent sources can be evaluated in three main criteria namely are timeliness as how fast the reaction to the security threat as the accuracy of detecting these threat and how reliable the system in performing the necessary related defense action.

3.1 Development cycle of threat intelligence

It is important for cyber security analysts conducting threat intelligence to adopt best practices methodology in developing their threat intelligent techniques and solutions [15]. This Intelligence cycle can be defined by five phases. In the first phase of requirement, the security intelligence professionals get information from their top management about what type of information they should gather and this information usually are considered the facts as main concerns of the end-clients or customers of the organization. Analysis is the second phase of the intelligence cycle with the purpose of turning these collected facts into actionable intelligence. For example, intrusion detection log files can be collected and analyzed in response to the rise of SSH attacks. For the sake of informative decision-taking procedures, the third phase of intelligence as dissemination to share useful information to end-clients in forms of technical reporting. Finally, feedback from end-client should be gathered to determine their satisfaction level and how to improve intelligence collection efforts in near future [16].

3.2 Threat indicators tools management

Threat indictors are the properties that describe a threat. These types of information are used to identify and describe certain threat. These indicators can include IP addresses, signatures of malicious detected files, highlighted communication patterns and any other types of identifiers can be used in cybersecurity to threat intelligence as all these collaborators should understand certain common communication language. So, if threat detected with the aids of these tools it will be very easy and efficient to inform other about that particular threat even in automated fashion [17]. The fist important threat sharing indicator tool for information sharing is the Cyber Observable Expression (CybOX) for categorizing security observations that helps in understand the properties of the intrusion attempts or malicious software. The second tool for this information exchange is the Structured Threat Information

Expression (or STIX) which is a standardized language that communicate security information between organizations and their systems. It uses the properties from the first tool and make language easy to use in structured manner. Trusted Automated Exchange of Indicator Information (TAXII) is tool containing services to effectively share security information between organizations and their systems. So at the end. TAXII can considered as framework for exchanging messages written as STIX language [18].

3.3 Threat intelligent information sharing

Previously discussed technological tools as TAXII, STIX and CybOX are enablers for sharing threat intelligence information between different organizations and systems. These tools give and added security values to different business functions within an organization such as the incident response team, vulnerability management team, risk management team, security engineering team and detection and monitoring team. The automated share information between tools and these different teams is the key achievement issue here. Threat information sharing in collaboration manner between different organizations is highly recommended and required and to facilitate such information sharing, Information and Analysis Centers (or ISACS) bring together different cybersecurity teams from different organizations to help sharing security of specific industry in confidential manners [19].

3.4 Use cases where threat intelligent is highly effective solution

Some of the use cases which show the importance of threats intelligence development in detection and treatments for these security problems can be shown for detecting unauthorized network connections, monitoring events that may change the user credentials, monitoring antivirus logs to identify insecure ports and services, managing replicas to ensure data protection and generating compliance reports in suitable formats by collecting system and security logs. Security compliance as GDPR can be enhanced significantly with good utilization of threat intelligent modeling for use cases where data protection is needed as verification and auditing security control, enable periodical reporting to data owners by providing structural access to log information, monitoring critical changes of users credentials also, managing data breaches by managing security alerts and analyze the full impact of the incidents.

4. Threat modeling

Threat modeling to be performed effectively, it needs some preprocessing modules such as performing threat research and then identifying and understanding the different types of possible threats [20, 21]. Once an organization can effectively model possible threats then threat hunting can be performed to significantly reduce and manage the different cybersecurity threats.

4.1 Conducting threat research

This step is important to understand better the environment that certain organization operate in also, to understand the motivations and levels of capabilities of the potential attackers. This can lead to better understanding of how to defend against these possible attacks. The aim of threat research is to know how attackers think and behave. Two important techniques can be followed in threat research to identify potential threats. The first technique is the reputational threat research to discover potential attackers based on their IP addresses, emails, domains that been previously involved in some attacks. This is very good practice to block future possible attacks from these sources. The second technique is the behavioral threat research aiming to potential malicious actors by observing the similarities of their behaviors when they attacked in the past [22].

4.2 Threat identification and understand the different types of attacks

To help organization keep tracking different types of threats efficiently, it is highly recommended that the security professionals use threats modeling techniques that can classify the different potential threats and categorize them based on their degree of risks. To properly identify the potential threat in an organization, a structured approach as threat management can be used. This structured approach can be utilized in three ways to efficiently identify threats. The first is the assets-focused approach and here the analysts base their analysis on an organization asset's inventory to identify the potential threat for each asset. The second structure approach in analyzing possible threat is the threat focused approach to properly understand all the possible threat that might affect the different information system within the organization. As for example, the different hacking techniques that might gain access to the network. These type of hacking can come from different parties include known hackers, trusted partners and even from the employees. Finally; a service focused approach can be used to identify the impact of various threats on each specific service when different services in an organization offered by different service providers. For example, when an organization is using certain API and expose it to the public, it is good practice to think about all the interfaces offering by that API and the threat can be associated with each interface. Identifying properly the different threat an organization maybe can be threaten by, is the first step toward proper threat modeling processes [23].

Once different security threats can be properly identified, security analysts should move forward to fully understand the possible attacks. The most commonly used model that help in categorizing these attack is the Microsoft STRIDE model. In this model. Each letter represents a category of attack as S stand for spoofing attack which uses falsified identity information to get access to the system and here the best control against spoofing is to use strong authentication. T indicate Tempering attack which is type of attack that make unauthorized changes to the system and disrupt the data integrity. R indicates Repudiation which is type of attack aiming to deny responsibility for an action and even can go further in blaming third-party, here digital signature can be very useful against such type of attack. I indicate information disclosure and in this type of attack, a theft of confidential information is intended and disclose it publically. D refers to the denial of service attacks (DOS) and this attack is trying to prevent the legitimate users accesses to information or the system they need. Finally; E standing for elevation of privileges which is also sometimes knowns as privileges escalation. This attack tries to use normal user account and then transform that to superuser account or root account in a purpose to exceed legitimate privileges [24]. A system diagrams that illustrate the data flow and relations between system modules is quite helpful in understanding the impact of different attacks in certain organization. These types of diagrams can be used in reduction analysis that breaks down the system into smaller components to properly perform assessment in each of them. This helps in simplifying complex systems to make thorough security reviews.

There are two important terms that should be clear to the security analysts as the "Total Attack Surface" which considers all of the systems and services that could be considered as potential entry points for an attacker. Also, the "Attack Vector" which can be defined as a means used by an attacker to gain initial access to a system or network [25].

4.3 Threats modeling as threat risk management

Threat modeling involve some important factors such as capabilities of the malicious hackers. Here by understanding the levels of sophistication and tools available to the potential hackers, it can give better understanding about how these attackers may approach and attack an organization. Another factor to be understood is the total attack surface and the potential attack vectors as these are two keys of characteristics to understand the types of attacks to be faced. Then, the factor of Impact as prioritize the different types of threats. Finally; the factor of Likelihood as combination between the impact of a threat can cause in an organization if it occurred and the likelihood of that threat to be materialized. As recommendation of the adoption of best practices, the threat modeling should be periodically prompt analysis of the security infrastructure. The significant benefit of using the efficient threat modeling is that it can detect repeated system inefficiencies such as data theft or data leakage and that may indicate the importance in using for example a data loss prevention (DLP) system to help cover the inefficiencies [26].

4.4 Effective threat hunting after proper threats Modeling

Threat hunting is an organized and systematic approach to clearly discover and find indicators of compromise on networks using different analytical techniques [27]. Threat hunting uses a combination tested security techniques as well as new analytic tools and technologies to monitor and tack signs of suspicious activities. Google trends can be considered as very good example in this context and it shows us how threat hunting grew rapidly recently as organization adopted this new approach. Threat hunting requires mind shifting from defense-focused to offense-focused approach. Here is very good to think as hackers who involved on activities attacking our organization. To conduct effective threat hunting, it is highly recommended to begin by establishing hypotheses and these hypotheses can be based upon profiling threat actors and their engaged activities or maybe hypotheses can be formulated based on possible information vulnerabilities. Once these hypotheses formulated then the thinking should be focused on the indicators of compromises that can be associated to these hypotheses if we assume them are true. These indicators might be considered as any unusual signs in the system such as unusual binary files with malicious or unknown content or some unpredicted modifications appeared in the system. This may include as well unexpected processes in the system or pattern of unusual consumption of resources. Sometimes even presence of unexpected account can be pointing or indicating to possible intended attack. Deviation of network traffic patterns is also considering obvious indicator here. Unexplained log entries and unapproved system configurations changes. All these indicators are the core of the threat hunting process [28].

4.5 Overall advantages and limitations of threat modeling

Six main beneficial aspects can be explained as on how threat modeling can help the security team to significantly enhance safety of the organization's technology assets. These benefits can be explained as the following points:

- Reducing attack surface: the attack surface here can be considered as the total number of vulnerabilities that an organization might be exposed to across the entire enterprise environment. This can be achieved as the ability to identify, track and maintain list of vulnerabilities to help security team take the important steps to mitigate them. Also, reducing the attack complexity is another very important benefit of adopting threat modeling in reducing attack surface. This can be accomplished by helping the team to breakdown a system and look at it from different perspectives to better understand it from end to end and this can help a lot in preventing risks to be propagated into end line. In this context, it is worthy to mention also that threat modeling in reducing the area of exposure and minimizing the attack surface of a system.

- Prioritizing threads, mitigation efforts and budgeting: threats modeling help organization quantifying risks and vulnerabilities and focus their attention and resources to minimize surface attacks in purposeful and effectives ways.

- Identify and eliminate single point of failure: organizations adopting a layered view of defensive tools to protect their assets can gain the advantage of reducing the chance that allow cyber attackers to take the maximum benefit of a single point of failure in a system.

- Understand the complete cyber kill chain: the kill chain breaks down the security individual steps and tactics, then evaluate and test for risks and communicate for each of them. This is in the line of the efforts to allow organizations to stop security threats at each stage.

- Improve organization's security maturity: by quantifying existing security practices, monitor security adopted programs and better structuring of security evaluation standards and policies.

- Improve organizational security posture at the individual application level: the aspects of achieving this can be highlighted as increase operational feasibility as developers can focus their attention on developing fixes and innovative service while security experts ensure solid controls are in place. Also, quality assurance can be more guaranteed at the early stage of the system which is the design phase as the key security mitigations can be considered as secure coding guidelines. Threat modeling also improves collaboration by mixing perspectives and experiences of different security professional teams.

Although threat modeling helps in many aspects in stopping and mitigating the risks of security breaches and attacks, still there are considerable challenges to be highlighted for the wide practical deployment and adoption of threat modeling. These challenges can be identified in five main aspects as the following:

Challenge 1: Processes saturation related to Threat Modeling

This challenge is due to numerous availability of threat modeling methodologies that may create confusion especially for teams lacking highly security expertise. This may lead to wrong choices for defense policies or cybersecurity investments.

Challenge 2: Scaled-up Modern Application Deployment

The recent departure of IT applications from physical servers and networking infrastructure to cloud computing infrastructure added new complexities related to responsibilities, expanded technologies, scope changes and associated risks which are no usually easy to be handled by the development teams.

Challenge 3: More systems entries points that are still not well recognized

The most obvious examples here are modern cloud services provider like AWS where many entry points are not yet recognized and these may include publicly-exposed management plane, APIs and services. These are significantly more complex in comparison with known entry points especially for the Data Flow Diagram (DFDs) and Processes Flow Diagram (PFDs).

Challenge 4: Vulnerability May Be Raised in Modern Authentication Security Token

As an example here, in AWS, temporary authentication tokens are transferable and they might be used outside the application environments. This is for sure can create a new security threat to be considered.

Challenge 5: Difficulties in Breaking Down Threats and Well-understanding of the Actual Risk

In some cases, it is difficult to determine the high-level threads and then breaking them to sub threats for easier deal with them. Also, it is challenging sometimes to identify the failure conditions in the system that may leads to threads. A deeper understand of these conditions is always preferable to the efforts to understand and mitigate these risks. Security teams should have the right framework and techniques for robust application security to effectively predict future and possible attack scenarios.

5. Environmental reconnaissance

Reconnaissance can be defined in the context of cybersecurity as the practice of discovering and collecting information about a system to facilitate the activities of attackers. There are many tools that can be used for reconnaissance as the following section explain them under related categories [29].

5.1 Social engineering techniques as reconnaissance tools

These types of attacks can be used by attackers as psychological tricks to manipulate people and pushing them to do certain actions as exposing some sensitive information that can be seriously harmful for organization's security [30]. Good example of that, when attackers pretending to act a help desk technician and attempting to trick user into revealing his/her password on a telephone call. Basically, social engineering attacks can be understood as online running a con and there are six main reasons that make social engineering attacks feasible. They are authority and trust, intimidation, consensus and social proof also scarcity, urgency and familiarity and liking. In the first type of authority and trust, users can be tricked as they welling to follow orders from an unauthorized person due to perceived authority and assumption of trusts. In the second reason that making social engineering attacks feasible is by pushing people to do things they supposed not to do by scaring people as telling something bad will be happened to them or to their organizations. In the third social engineering tactic namely as consensus and social proof, individuals are not exactly knowing how to react to certain situation and they just look to others and follow their behaviors. Sometimes this is called the herb mentality. Scarcity tactic can be achieved by making people believe that they will be missed-out if they not act quickly in certain scenario. For example, attackers can push people to install unauthorized Wi-Fi router in the office as the attackers claim that they upgrading Wi-Fi existing technology and the newly brand technology has left only one router at the time [31]. In urgency, the attackers create pressure environment on people to push them in this situation to act quickly as the time is running-out. In the final tactic as familiarity or liking, the social engineer, use flattery, false compliments or even fake relationships to manipulate the target's good side and then influence their activities.

As a conclusion, if you are cybersecurity analyst, you should be aware that attackers can use different social engineering attacks against your organization as attempts to gather critical information and influence activities.

5.2 DNS harvesting as reconnaissance tools

In general, Domains names and their associated IP addresses can be considered an excellent starting point to gather useful information about the true owners of systems. There are some utilities that can be used to learn more about remote systems. The first thing can be considered here is trying to learn about the host behind certain domain name. Here it is always useful to remember that the DNS translates domain names to IP addresses [32]. It is interesting point to know that, usually we can perform lookups functions manually to find out certain IP addresses associated with their domain name. To perform domain lookups on Linux or Mac systems, the dig command is the primary tool here. The alternative in windows systems is the nslookup command and it works basically in the same way. In some cases, where the IP address may consider the source of suspicious log entries or a host that might be shown in a netstat command. In general consideration the IP address or domain name that needed to learn more about, the whois utility allows to know more about the ownership of particular domain name or IP address. This whois lookup utility can be offered in many web sites such as domaintools.com. This site can give a good information about certain domain name such as the registrant organization and through which DNS registrar, when it was created and renewed. Even these websites utility can give the contact information for the owner of that particular domain such as the e-mail address, the street address and the telephone number in case needed to communicated with them regarding this domain. The same utility website can be used for looking up certain IP address and then can get all the information about that IP address as to which domain this IP address is registered for and the contact information that can be used to contact that organization when that is needed [33].

A very useful reconnaissance too can be considered also here in this context is Reverse Whois Lookup which allows to determine all domain names related to an email address. This can be very helpful to understand how different domains name may be related to each other. In general, there is a wide variety of Reverse Whois tools available on internet and the good example here is viewdns.info and this can be very useful to those attackers engaging in an attacks on particular domains as it gives good ideas about the owners and what other domain they maybe own [34].

As a conclusion, it is always recommended for all cyber security analyst professional to use all these tools and techniques against their own IP addresses and domains names to learn what things potential attackers might discover about the organization.

5.3 Network scanning and mapping

Discovering networks topologies and how these networks are connected to the hosts. Also, discovering the open ports of communication in certain server and the running operating system fingerprint are considered the most important types of information that hacker look for when attacking an organization network. The most important tools they can use here are the NMAP and ZENMAP [35]. These tools help in identifying the connected hosts and the topology of that connected network also, the discovery of open service port number and the Server OS running and its version. Here for example, these tools can show a report telling that there is a server running and listening to the port 3389 and here it can be discovered that this port is used by Microsoft Remote Desktop service. As difference between NMAP and ZENMAP, it is worthy to mention here that ZENMAP is extended graphical capabilities where graphical representation of the network topology can be presented with the capability to focus on certain host and analyze it in term of running port services as well as OS fingerprinting [36].

5.4 Passive and active enumeration tools

Passive enumeration tools such as Wireshark can gather information about network without directly interacting with the network or announcing their presence. In the other hand, active enumeration tools are directly interacting with the system to be able to capture more complete information but here there is a risk of being discovered by the system administrator. As example, NMAP conducts port scanning by sending requests to the remote server so this tool can be considered as active enumeration. Another interesting example here is Hping, which allows to scan specific TCP port such as port 80 that is used for HTTP connection or port 43 used for HTTPS or port 22 for secure shell protocol [36]. Using this Hping tool we can determine the level of security configurations and the potential security vulnerabilities in the system. Hping considered to be very useful tool as it allows of customization of the content of the sent packet for the purpose of advance penetration testing. Another interesting enumeration tool which is considered as opportunistic python script, is Responder. This tool waits for broadcast requests and then response to them for that it is called opportunistic tool as it captures traffic intended for other system aiming to trick users and drag them to log into a fake server then, Responder can be able to capture the user's credentials to be used later in other attacks [37, 38].

5.5 Protocols analyzer tools

Protocol analyzers tools are important for both professionals of network analyzing and Cyber security as well. These tools have the capability to capture actual packet traveling on a network and investigate them in great details. Wire Shark is the most famous tool under this category. It is free and open source packet analyzer that can be used for network troubleshooting based on the actual network traffic. The utilization of this tool for troubleshooting may include dropped packets, latency issues and discovery of unusual traffic based on some malicious activities [39, 40].

5.6 Tools for wireless reconnaissance

These types of tools usually analyze the wireless network environments and used mainly to test the security of a wireless network. The most common toolkit here is the Aircack-ng. This toolkit is a collection of tools that can be used in different stages in the context of the wireless reconnaissance efforts. In this section, the core component of this toolkit that can be used to test the security of the wireless networks. The first tool is Airmon-ng puts wireless interfaces into promiscuous mode to enable eavesdropping on wireless traffic. Airodump-ng is a tool for wireless packet capturing to capture data over a wireless network. Aircrack-ng is the most important tool in this toolkit that come with the capabilities to break encrypted keys used in the wireless network. Airreply-ng is tool that is used for traffic injection that can be used to inject certain traffic into a wireless network. This extra injected traffic can be reached to level enough to force the disconnect of legitimate wireless clients [41].

Another two widely used wireless reconnaissance tools are namely Reaver and Hashcat. Reaver is a tool allowing of Wi-Fi password retrieval as it exploit Wi-Fi Protected Setup (WPS) exposing the vulnerabilities to retrieve WPA and WPA2 passwords (Ram, 2016). As a good security practice, it is recommended to disable WPS encryption in the Wi-Fi network to not be vulnerable to Reaver attack. Hashcat which can be used to conduct Brute Force attacks against hashed passwords including WPA and WPA2 passwords [42, 43].

5.7 The different perspectives of network security reconnaissance

It's extremely important to understand the perspectives that can viewed on a network when conducting network reconnaissance. The different levels of network security controls can significantly limit accesses as well as the accuracy of the network scans. Network devices such as firewalls, intrusion detection system, switches and routers, all these devices restricted the access to the resources of the network and at the same time limit the accuracy of network scanning [44]. In line of that, open ports numbers can be seen differently if the scan of the network is performing internally or externally as the restricted firewall rules can be applied differently here. In brief, running scan internally or externally from the network can provide different views of the network as externally give the cybersecurity analyst with attack potentially from the internet while internally provides view of the internal attacks and both attacks aspects are important information. Besides these internal and external aspects of the network scanning and security analysis, there are other aspects to consider in network security reconnaissance such as if the scanned system is physical servers or virtual servers, are these servers running on-promises or in clouds? Are they wired or wireless? These all are affecting the scanning results which in turn affect the security reconnaissance aspects [45].

6. Cloud-based security services

In the past decade, cloud computing can be considered as the most transformative development in information technologies. Most of the organizations around the world are adapting and retooling their entire IT strategies to be integrated with the cloud. Cloud in its simplest definition is the delivery of computing resources/services to remote client over a network. Example of these services or resources can be explain as accessing the Gmail account is considered a use of cloud computing as google give an email service over the internet as no need for the end-user to know or care about the massive technical infrastructure that make the Gmail works. Also, when someone build a server in Amazon Web Services, he/she making use of cloud computing. Amazon make it appear as own server for that person but in fact, it's virtual server running in massive Amazon data center as hardware shared with many customers in the same time. The beauty of that is the technology that make this happening is invisible to the end-uses. Even, when writing scripts in SaleForce.com to automatically follow-up with clients, it is a use case of cloud computing as the written code of the scrip to make the follow-up e-mail happen is executed on top of Salesforec's cloud-based platform [46].

The discussion about cloud resources and services should cover the main cloud security risks. The risks associated with Application Programming Interfaces or APIs should be highly considered here. APIs provide developers with programmatic access to services as for example, Amazon Web Service's API enables creation and provisioning of server instances. If these APIs performed by unauthorized individuals, this will be considered a serious security risk issue. Developers should require strong authentication to prevent misuses or insecure APIs. Key management is important here and considered as same level importance for encryption keys as losing control of API key is a very serious risk issue.

The most common cloud-based security services of interests can be explained as the following sections.

6.1 Cloud identity and access management (IAM)

The threats and security challenges can be addressed under this category can include: Identity theft, unauthorized access, privilege escalation, Insider threat, fraud. Possible actions can be taken can be considered as assigning of duties based on identity entitlement and compliance-centric reporting [47–49].

6.2 Cloud data loss prevention

The threats and security challenges can be addressed under this category can include: Data misused by datacenter operator or others by unauthorized access, compromising the data integrity, issues caused by data sovereignty [50]. Possible actions can be taken for enhancement could be file/directory integrity via hashing, smart response for unstructured data matching and integrating intrusion detection solutions [51].

6.3 Cloud web security

The threats and security challenges can be addressed under this category can include: Malware, Spyware, Key loggers, Phishing, Viruses, Spams and Bandwidth consumption. Possible actions can be taken for enhancement could be: Policy enforcement to categorize web-sites security level, categorize websites based on IP/URL addresses, Domain rating and rating web-sites based on users' requests [52, 53].

6.4 Cloud E-mail security

The threats and security challenges can be addressed under this category can include: Phishing, Intrusion, Malware, Spam and address spoofing. Possible actions can be taken for enhancement could be: E-mail backup system policy, Data loss prevention for SMTP and webmail, Secure archiving, Mail encryption, Signing and time stamping [54, 55].

6.5 Cloud intrusion management

The threats and security challenges can be addressed under this category can include: Intrusion and Malware. Possible actions can be taken for enhancement could be: Centralized reporting with administrator notifications capability, prevent local evasion by remote storage or transmission of integrity information, AI anomaly detection and solid intrusion management of API [56, 57].

6.6 Cloud information security and event management

The threats and security challenges can be addressed under this category can include: Insecure APIs Interfaces, Malicious insiders, Account hijacking, Fraud. Possible actions can be taken for enhancement could be: Standardization of log formats, Log monitoring, Heuristic control and use of specialized systems, integration with physical security (cameras and phones) [58, 59].

7. Conclusions

Knowledge based system containing the different and latest types of cybersecurity threats is highly recommended to be integral parts of the operation in today IT organizations. This knowledge-based system can be extended to decision-support system to detect, prevent and minimize the risks of security risks in real-time manner. But, reaching to this level of security maturity level is never be an easy task as this requires the engagement and being updated about the latest practices of different threats identification, intelligent network traffic analysis, ethical hacking to probe the system under analysis for possible security attacks and risks management policies for system recovery if the attacks been successful. Having the capabilities for fast adaptation is the key point here to prevent and treat most of the outsiders or insider's security attacks. With all these effort cybersecurity intelligent threat modeling can be easily and widely adopted for solid security defense in modern IT information systems including cloud computing and in the context of agile software project management such as DevOps agility context.

Author details

Mohammed Saeed Jawad[1] and Mohammed Hlayel[2]*

1 FSKTM, CIAS Focus Research Group, UTHM University, Parit Raja, Johor, Malaysia

2 FSKTM, UTHM University, Parit Raja, Johor, Malaysia

*Address all correspondence to: mo.hlayel@gmail.com

References

[1] Craigen D, Diakun-Thibault N, Purse R. Defining cybersecurity. Technology Innovation Management Review. 2014;**4**(10):13-21

[2] Roldán-Molina G, Almache-Cueva M, Silva-Rabadão C, Yevseyeva I, Basto-Fernandes V. A comparison of cybersecurity risk analysis tools. Procedia Computer Science. 2017;**121**:568-575

[3] Rehman R, Hazarika GC, Chetia G. Malware threats and mitigation strategies: A survey. Journal of Theoretical and Applied Information Technology. 2011;**29**(2):69-73

[4] Chen Z, Roussopoulos M, Liang Z, Zhang Y, Chen Z, Delis A. Malware characteristics and threats on the internet ecosystem. Journal of Systems and Software. 2012;**85**(7):1650-1672

[5] Aycock J. Getting There. In: Spyware and Adware. Advances in Information Security. Boston, MA: Springer; 2011;**50**. https://doi.org/10.1007/978-0-387-77741-2_2

[6] Yadav N, Kaur G, Kaur S, Vashisth A, Rohith C. A complete study on malware types and detecting ransomware using API calls. In: 2021 9th International Conference on Reliability, Infocom Technologies and Optimization (Trends and Future Directions) (ICRITO). 2021. pp. 1-5

[7] Bansal U. A review on ransomware attack. In: 2021 2nd International Conference on Secure Cyber Computing and Communications (ICSCCC). 2021. pp. 221-226

[8] Shalaginov A, Dyrkolbotn GO, Alazab M. Review of the malware categorization in the era of changing cybethreats landscape: Common approaches, challenges and future needs. In: Malware Analysis Using Artificial Intelligence and Deep Learning. Cham: Springer; 2021. pp. 71-96

[9] van Oorschot PC. Malicious Software. In: Computer Security and the Internet. Cham: Springer; 2021. pp. 183-211

[10] Yongwang T, Xin L, Qizheng D. Malicious Code Detection Technology based on Bi-GRU and Self-attention. In: Proceedings of the 2019 2nd International Conference on Algorithms, Computing and Artificial Intelligence. 2019. pp. 585-590

[11] Sharma S, Rama Krishna C, Sahay SK. Detection of advanced malware by machine learning techniques. In: Ray K, Sharma T, Rawat S, Saini R, Bandyopadhyay A, editor. Soft Computing: Theories and Applications. Advances in Intelligent Systems and Computing. vol 742. Singapore: Springer; 2019. https://doi.org/10.1007/978-981-13-0589-4_31

[12] Subedi KP. PhD Dissertation. A Framework for Analyzing Advanced Malware and Software. The University of Memphis. 2018

[13] Conti M, Dargahi T, Dehghantanha A. Cyber threat intelligence: Challenges and opportunities. In: Cyber Threat Intelligence. Cham: Springer; 2018. pp. 1-6

[14] Tounsi W, Rais H. A survey on technical threat intelligence in the age of sophisticated cyber attacks. Computers &Security. 2018;**72**:212-233

[15] Abu MS, Selamat SR, Ariffin A, Yusof R. Cyber threat intelligence–issue and challenges. Indonesian Journal of Electrical Engineering and Computer Science. 2018;**10**(1):371-379

[16] Du L, Fan Y, Zhang L, Wang L, Sun T. A summary of the development of cyber security threat intelligence sharing. International Journal of Digital Crime and Forensics (IJDCF). 2020;**12**(4):54-67

[17] Brown S, Gommers J, Serrano O. From cyber security information sharing to threat management. In: Proceedings of the 2nd ACM Workshop on Information Sharing and Collaborative Security. 2015. pp. 43-49

[18] Casey E, Back G, Barnum S. Leveraging CybOX™ to standardize representation and exchange of digital forensic information. Digital Investigation. 2015;**12**:S102-S110

[19] Abomhara M, Køien GM. Cyber security and the internet of things: Vulnerabilities, threats, intruders and attacks. Journal of Cyber Security and Mobility. 2015;**4**:65-88

[20] Hussain S, Kamal A, Ahmad S, Rasool G, Iqbal S. Threat modelling methodologies: A survey. Science International (Lahore). 2014;**26**(4):1607-1609

[21] Bojanc R, Jerman-Blažič B. A quantitative model for information-security risk management. Engineering Management Journal. 2013;**25**(2):25-37

[22] Stedmon A, Paul D. Conducting ethical research in sensitive security domains: Understanding threats and the importance of building trust. In: Iphofen R, O'Mathúna D, editors. Ethical Issues in Covert, Security and Surveillance Research (Advances in Research Ethics and Integrity, Vol. 8), Emerald Publishing Limited. Bingley. 2021. pp. 159-176. https://doi.org/10.1108/S2398-601820210000008012

[23] Agrafiotis I, Nurse JR, Goldsmith M, Creese S, Upton D. A taxonomy of cyber-harms: Defining the impacts of cyber-attacks and understanding how they propagate. Journal of Cybersecurity. 2018;**4**(1):tyy006

[24] Hamed T, Ernst JB, Kremer SC. A survey and taxonomy of classifiers of intrusion detection systems. In: Computer and Network Security Essentials. Cham: Springer; 2018. pp. 21-39

[25] Simmons C, Ellis C, Shiva S, Dasgupta D, Wu Q. AVOIDIT: A cyber attack taxonomy. In: 9th Annual Symposium on Information Assurance. 2014. pp. 2-12

[26] Alneyadi S, Sithirasenan E, Muthukkumarasamy V. A survey on data leakage prevention systems. Journal of Network and Computer Applications. 2016;**62**:137-152

[27] Steingartner W, Galinec D, Kozina A. Threat defense: Cyber deception approach and education for resilience in hybrid threats model. Symmetry. 2021;**13**(4):597

[28] Buchanan B. The Cybersecurity Dilemma: Hacking, Trust and Fear Between Nations. Oxford University Press. 2017. Retrieved 19 Jun. 2022, from https://oxford.universitypressscholarship.com/view/10.1093/acprof:oso/9780190665012.001.0001/acprof-9780190665012

[29] Arabia-Obedoza MR, Rodriguez G, Johnston A, Salahdine F, Kaabouch N. Social engineering attacks a reconnaissance synthesis analysis. In: 2020 11th IEEE Annual Ubiquitous Computing, Electronics & Mobile Communication Conference (UEMCON). IEEE; 2020. pp. 0843-0848

[30] Ozkaya E. Learn Social Engineering: Learn the art of human hacking with an internationally renowned expert. Packt Publishing Ltd. 2018

[31] Fiermonte M. The Threat of Social Engineering to Networked Systems. Utica College; 2019

[32] Hu Q, Asghar MR, Brownlee N. Measuring IPv6 DNS reconnaissance attacks and preventing them using DNS Guard. In: 2018 48th Annual IEEE/IFIP International Conference on Dependable Systems and Networks (DSN). 2018. pp. 350-361

[33] Skwarek M, Korczynski M, Mazurczyk W, Duda A. Characterizing vulnerability of DNS AXFR transfers with global-scale scanning. In: 2019 IEEE Security and Privacy Workshops (SPW). IEEE; 2019. pp. 193-198

[34] Hudák P. Analysis of DNS in Cybersecurity. Brno: Masaryk University, Faculty of Informatics; 2017

[35] Calderon P. Nmap: Network Exploration and Security Auditing Cookbook - Second Edition (2nd ed.). Packt Publishing. 2017. Retrieved from: https://www.perlego.com/book/527158/nmap-network-exploration-and-security-auditing-cookbook-second-edition-pdf (Original work published 2017)

[36] Lastovicka M, Jirsik T, Celeda P, Spacek S, Filakovsky D. Passive os fingerprinting methods in the jungle of wireless networks. In: NOMS 2018-2018 IEEE/IFIP Network Operations and Management Symposium. 2018. pp. 1-9

[37] Bhatnagar D, Som S, Khatri SK. Advance persistant threat and cyber spying-the big picture, its tools, attack vectors and countermeasures. In: 2019 Amity International Conference on Artificial Intelligence (AICAI). IEEE; 2019. pp. 828-839

[38] Ramadhan RA, Aresta RM, Hariyadi D. Sudomy: Information gathering tools for subdomain enumeration and analysis. In: IOP Conference Series: Materials Science and Engineering. 2020

[39] Bagyalakshmi G, Rajkumar G, Arunkumar N, Easwaran M, Narasimhan K, Elamaran V, et al. Network vulnerability analysis on brain signal/image databases using Nmap and Wireshark tools. IEEE Access. 2018;**6**:57144-57151

[40] Sija BD, Goo Y-H, Shim K-S, Hasanova H, Kim MS. A survey of automatic protocol reverse engineering approaches, methods, and tools on the inputs and outputs view. Security and Communication Networks. 2018;2018:17, 8370341. https://doi.org/10.1155/2018/8370341

[41] Astudillo K. Wireless Hacking 101 ([edition unavailable]). Babelcube Inc. 2021. Retrieved from: https://www.perlego.com/book/2984611/wireless-hacking-101-pdf (Original work published 2021)

[42] Ram JR Sak B. Mastering Kali Linux Wireless Pentesting [Book]. Publisher: Packt Publishing Ltd. 2016. 310 p

[43] Lundgren M, Persson J. Constructing and Evaluating a Raspberry Pi Penetration Testing/Digital Forensics Reconnaissance Tool (Dissertation). 2020. Retrieved from http://urn.kb.se/resolve?urn=urn:nbn:se:hh:diva-42805

[44] Mazurczyk W, Caviglione L. Cyber reconnaissance techniques. Communications of the ACM. 2021;**64**(3):86-95

[45] White R, Caiazza G, Jiang C, Ou X, Yang Z, Cortesi A, et al. Network reconnaissance and vulnerability excavation of secure DDS systems. In: 2019 IEEE European Symposium on Security and Privacy Workshops (EuroS&PW). 2019. pp. 57-66

[46] Manna M. A cloud-based encryption for document storage using salesforce. com. Journal of Engineering and Applied Science. 2018;**13**:2382-2387

[47] Pramod N, Muppalla AK, Srinivasa KG. Limitations and challenges in cloud-based applications development. In: Software Engineering Frameworks for the Cloud Computing Paradigm. London: Springer; 2013. pp. 55-75

[48] Indu I, Anand PR, Bhaskar V. Identity and access management in cloud environment: Mechanisms and challenges. Engineering Science and Technology: An International Journal. 2018;**21**(4):574-588

[49] Schulze R. Identity and access management for cloud services used by the payment card industry. In: International Conference on Cloud Computing. Cham: Springer; 2018. pp. 206-218

[50] Manzoor CS, Shabina G. Challenges of data protection and security in cloud computing. Proceedings of the International Conference on Innovative Computing & Communication (ICICC) 2021. July 3, 2021. Available at SSRN: https://ssrn.com/ abstract=3879599 or http://dx.doi. org/10.2139/ssrn.3879599

[51] Bhardwaj A, Goundar S. A framework to define the relationship between cyber security and cloud performance. Computer Fraud & Security. 2019;**2019**(2):12-19

[52] Prasath R, Santhosh GT, Ratchnayaraj IAJ, Jemiline E. The security in web application of cloud and IoT service. Materials Today: Proceedings. 2020

[53] Paul P, Aithal PS. Cloud security: An overview and current trend. International Journal of Applied Engineering and Management Letters (IJAEML). 2019;**3**(2):53-58

[54] Hashmi A, Ranjan A, Anand A. Security and compliance management in cloud computing. International Journal of Advanced Studies in Computers, Science and Engineering. 2018;**7**(1):47-54

[55] Cidon A, Gavish L, Bleier I, Korshun N, Schweighauser M, Tsitkin A. High precision detection of business email compromise. In: 28th {USENIX} Security Symposium ({USENIX} Security 19). 2019. pp. 1291-1307

[56] Helmiawan MA, Fadil I, Sofiyan Y, Firmansyah E. Security model using intrusion detection system on cloud computing security management. In: 2021 9th International Conference on Cyber and IT Service Management (CITSM). 2021. pp. 1-5

[57] Devi S, Sharma AK. Understanding of intrusion detection system for cloud computing with networking system. International Journal of Computer Science and Mobile Computing (IJCSMC). 2020

[58] Adam I, Ping J. Framework for security event management in 5G. In: Proceedings of the 13th International Conference on Availability, Reliability and Security. 2018. pp. 1-7

[59] Al-Rashdi ZA, Dick M, Al-Rashdi RA, Al-Husaini Y. Information Security Accountability in the Cloud Computing Context—A Comprehensive Review. In: Montasari R, Jahankhani H, Al-Khateeb H, editors. Challenges in the IoT and Smart Environments. Advanced Sciences and Technologies for Security Applications. Cham: Springer; 2021. https:// doi. org/10.1007/978-3-030-87166-6_8

Chapter 6

Advance in Keyless Cryptography

Valery Korzhik, Vladimir Starostin, Victor Yakovlev, Muaed Kabardov, Andrej Krasov and Sergey Adadurov

Abstract

The term "keyless cryptography" as it is commonly adopted, applies to secure message transmission either directly without any key distribution in advance or as key sharing protocol between communicating users, based on physical layer security, before ordinary encryption/decryption procedures. In the current chapter the results are presented concerning to keyless cryptography that have been obtained by authors recently. Firstly Shamir's protocol of secure communication is considered where commutative encryption procedure is executed. It has been found out which of the public key algorithms can be used with such protocol. Next item of consideration concerns Dean's and Goldsmith's cryptosystem based on multiple-input, multiple-output (MIMO) technology. It has been established under which conditions this cryptosystem is in fact secure. The third example under consideration is EVSkey scheme proposed recently by D. Qin and Z. Ding. It has been proven that such key distribution method is in fact insecure, in spite of the authors' claims. Our main result is a description of a key sharing protocol executing over public noiseless channels (like internet) that provides a key sharing reliability and security without any cryptographic assumptions.

Keywords: keyless cryptography, protocols, public cryptosystems, MIMO technology, cryptographic assumptions

1. Introduction

The term keyless cryptography dates back many years ago. See for example the paper [1]. One author of this chapter used also before this term many times, for example in the papers [2–4]. This term is commonly used in two senses: either in the scenarios where information transmission security is provided by special channel properties without execution of key sharing protocol in advance, or in the second scenarios where specific channel properties are used on the stage of secret key sharing between users, whereas later are executed ordinary key-based cryptography. An approach to a providing of security at the cost of channel properties was termed as physical layer security [5]. As a rule, it means that communication channel plays the role of some randomizer due to their specific stochastic properties. However, sometimes such randomization is provided by the use of artificial noises by the communicating users.

A natural question arises—why it is not sufficiently to execute algorithms of the so-called *public key cryptosystems* (PKC), invented by W. Diffie and M. Hellman and developed later by many cryptographers: A. Shamir, Rabin, El Gamal, McEliece, J. Massey, and others (see [6, 7])? The point is that PKC have some drawbacks that limit their practical use. The main of these are:

1. PKC are based on some cryptographic assumptions (integer factoring, discrete log computation, error correction by random liner codes [6]). Unfortunately, the most of such hard computational problems can be solved in polynomial time by quantum computers (QC) [8]. Although a practical implementation of such QC's is still highly conjectural [9], it would be a hazardous to ignore such attacks in the future.

2. The use of PKC is needed in an assistance of certificated center that would be guaranty an authenticity of public keys in order to prevent impersonation attack where an attacker camouflages itself by authorized user and could share with legitimate user falsified keys.

3. The third shortcoming is in a complexity of encryption/decryption algorithms because it requires performing operations with large integer values. It is especially important for mobile hardware devices.

One more way out to provide secure key sharing is an implementation of quantum cryptography [10]. But unfortunately such approach requires the use of very specific quantum devices.

The objections mentioned above give the reason to turn to keyless cryptography which is what we do in this chapter. In Section 2 we investigate Shamir's protocol for secure communication over public channels but without any key sharing in advance. This protocol corresponds to the first scenario. Section 3 presents Dean and Goldsmith cryptosystem that performs channel transmission like a randomizer in the first scenario. We turn out under which restrictions on the channel such protocol is in fact secure. In Section 4 the second scenario (with key sharing) is considered and it is shown that on the contrary to author's claims [11], such *key sharing protocol* (KSP) is in fact insecure. Section 5 corresponds also to the second scenarios. We present a new KSP and prove that such protocol for appropriated chosen parameters provides reliable shared keys which are secure in terms of Shannon information without any cryptographic assumptions. Section 6 summarized all previous results and presents our opinion regarding to possible future investigations.

2. Investigation of Shamir's protocol based on commutative cryptography

Shamir's protocol was described in the book [6]. This protocol can be executed for any cryptosystem (both symmetric and asymmetric) if they satisfy the following condition:

$$f_{K_A}\left(f_{K_B}(M)\right) = f_{K_B}\left(f_{K_A}(M)\right), \tag{1}$$

where $f_{K_A(K_B)}(M)$ is encryption algorithm using the keys K_A and K_B, respectively and M is any message. If relation (1) is true for some encryption algorithm $f_{K_A(K_B)}(M)$ then can be performed protocol presented in **Figure 1**.

If the relation (1) is valid, then we get:

$$C'_A = f_{K_A}^{-1}(C_B) = f_{K_A}^{-1}\left(f_{K_B}\left(f_{K_A}(M)\right)\right) = f_{K_A}^{-1}\left(f_{K_A}\left(f_{K_B}(M)\right)\right) = f_{K_B}(M) \quad (2)$$

On the next step user B decrypts $f_{K_B}(M)$ by the algorithm $f_{K_B}^{-1}(\cdot)$ and gets the message M. But the following question arises—for which cryptosystems the relation (1) is valid? Let us call such encryption algorithms, for which (1) is true, *commutative encryption* (CE). It is easy to verify that for such cryptographic standard as AES, 3DES, GOST [7] the relation (1) is not fulfilled for all messages. Let us consider public key cryptosystems and, first of all, Rivest-Shamir-Adleman system (RSA). It is well known that such cryptosystem is determined by the following parameters: p, q—different primes, $n = pq(modulo)$, $\varphi = (p-1)(q-1)$, e—encryption key, $1 \le e \le \varphi$, d—decryption key, $d = e^{-1} mod\ \varphi$, $\gcd(e, \varphi) = 1$, $1 \le M \le n-1$—integers. Encryption algorithm for RSA is $E = M^e mod\ n$; decryption algorithm is $M = E^d mod\ n$. Then for RSA condition (1) takes the form:

$$(M^{e_A} mod\ n)^{e_B} mod\ n = (M^{e_B} mod\ n)^{e_A} mod\ n \quad (3)$$

Since exponentiation by modulo n is a commutative operation, the relation (3) becomes trivially true. In fact

$$(M^{e_A} mod\ n)^{e_B} mod\ n = M^{e_A e_B} mod\ n = (M^{e_B} mod\ n)^{e_A} mod\ n = M^{e_B e_A} mod\ n.$$

On the one hand RSA system is self-sufficient one but on the other hand there may be situations where its public key e_B is unknown for the user A or should be secure. Then protocol shown in **Figure 1** could have a reason.

Example. Let $p = 3, q = 5, M = 2, e_A = 3, e_B = 5$. Then it is easy to verify that both parts of (3) are equal to 8.

Let us consider as next public key cryptosystem, Rabin's one, that is determined by the following parameters [7]: p, q are distinct primes (secret key), $n_{A(B)} = pq(modulo)$ is public key. $0 \le M \le n-1$, $E = M^2 mod\ n$. Then relation (1) takes the form:

$$\left(M^2 mod\ n_A\right)^2 mod\ n_B = \left(M^2 mod\ n_B\right)^2 mod\ n_A \quad (4)$$

It is easy to make sure that (4) is not satisfied, generally speaking. In fact, let us consider an example: $p_A = 3, q_A = 5, n_A = 15, p_B = 11, q_B = 7, n_B = 77, M = 5$.

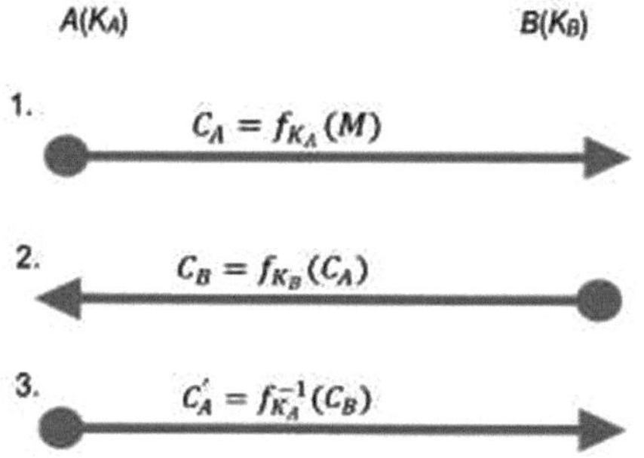

Figure 1.
Protocol based on relation (1) where $f_{K_A}^{-1}(\cdot)$ is decryption algorithm given the key K_A.

Then it is easy to check that the left side of (4) is equal to 23, while the right side is equal to 10.

In order to find out the reason of that fact, let us present both sides of (4) as the following values, respectively:

$$\begin{aligned}\left(M^2 mod\, n_A\right)^2 mod\, n_B &= \left(M^2 - n_A l\right)^2 mod\, n_B = \\ &= \left(M^4 - 2M^2 n_A l + n_A^2 l^2\right) mod\, n_B,\end{aligned} \tag{5}$$

where l is some integer.

$$\begin{aligned}\left(M^2 mod\, n_B\right)^2 mod\, n_A &= \left(M^2 - n_B m\right)^2 mod\, n_A = \\ &= \left(M^4 - 2M^2 n_B m + n_B^2 m^2\right) mod\, n_A,\end{aligned} \tag{6}$$

where m is some integer too. We can see that in general case $n_A \neq n_B$ the right sides of (5) and (6) are different values. It was mentioned in Section 1 that some PKC is vulnerable against QC's. Such of them that are resistant against QC's attacks were called *post quantum cryptosystems*. The most known among them is McEliece PKC [7]. The matrices that determine it are: $k \times n$ matrix $\tilde{G}_{A(B)} = S_{A(B)} G_{A(B)} P_{A(B)}$—public key where $S_{A(B)}$ is nonsingular random $k \times k$ matrix, $P_{A(B)}$ is permutation random $n \times n$ matrix, $G_{A(B)}$ is random Goppa code generating matrix. Matrices $S_{A(B)}$, $G_{A(B)}$, $P_{A(B)}$ are believed as secret key jointly with a binary random vector $Z_{A(B)}$ of the known length n, and with given weight $t_{A(B)}$. Encryption procedure is performed as follows:

$$E_{A(B)} = M\tilde{G}_{A(B)} \oplus Z_{A(B)}, \tag{7}$$

where " $\oplus$ " is bitwise modulo two addition. It follows from (7) that commutative property (1) for McEliece CS was looking as:

$$f_{K_A}\left(f_{K_B}(M)\right) = \left(M\tilde{G}_B \oplus Z_B\right)\tilde{G}_A \oplus Z_A = M\tilde{G}_B\tilde{G}_A \oplus Z_B\tilde{G}_A \oplus Z_A \tag{8}$$

$$f_{K_B}\left(f_{K_A}(M)\right) = \left(M\tilde{G}_A \oplus Z_A\right)\tilde{G}_B \oplus Z_B = M\tilde{G}_A\tilde{G}_B \oplus Z_A\tilde{G}_B \oplus Z_B \tag{9}$$

It is easy to see that the values in the right sides of (8) and (9) are, generally speaking, unequal one to another owing at least different vectors Z_A and Z_B. Hence McEliece CS does not correspond to CE algorithm. At a single glance stream cipher is looking as CE algorithm. In fact, the encryption procedure for this cipher is:

$$E(M, K) = M \oplus \gamma(K), \tag{10}$$

where $E(M, K)$ is a cipher text given secret key K, $\gamma(K)$ is an encrypting binary sequence (gamma) given the key K, " $\oplus$ " is bitwise modulo 2 addition, M is the binary message to be encrypted. It follows from (10) that the condition (1) takes the following form for γ-based stream ciphers:

$$M \oplus \gamma(K_A) \oplus \gamma(K_B) = M \oplus \gamma(K_B) \oplus \gamma(K_A) \tag{11}$$

that is valid trivially. The use of stream ciphers as CE in protocol shown in **Figure 1** was looking as it is presented in **Figure 2**.

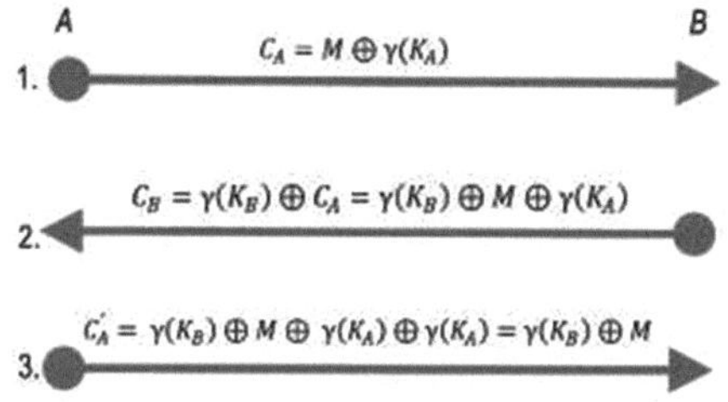

Figure 2.
Protocol of secret message transmission based on stream ciphers.

However we can see from **Figure 2** that eavesdropper receiving C_B and C_A' be able to bitwise add them by modulo 2 that gives the following binary sequence: $M \oplus \gamma(K_B) \oplus \gamma(K_B) \oplus M \oplus \gamma(KA) = \gamma(K_A)$. Next, attacker add bitwise modulo 2 the last sequence $\gamma(K_A)$ with the sequence C_A intercepted during the first transmission session that gives $M \oplus \gamma(K_A) \oplus \gamma(K_A) = M$ that is exactly the desired open sequence M. (It is worth to note that similar conclusion was mentioned in [6].)

3. Protocol of keyless Dean and Goldsmith system elaborated for secure message transmission through fading channels with executing MIMO technology

Cryptosystem proposed by Dean and Goldsmith in [12] belongs also to the class of keyless cryptography because it may provide a security of message transmission without a secret key sharing in advance. The main difference in the knowledge of legitimate users and eavesdroppers is only their different locations in space. But such cryptosystems can be used not in all possible scenarios but only in those ones where messages are transmitted over fading channels and with the use of a *massive multiple-input, multiple-output* (MIMO) technology.

For brevity, we denote such cryptosystem by abbreviation DGC. First, we consider the model with the main steps of its implementation for the particular case where the number of legal user receiving antennas n_r and eavesdropper antennas n_r' are equal to each other. (Later we consider more general case $n_r \neq n_r'$).

Legitimate channel between Alice (A) and Bob (B) is described by equation:

$$z = Ay + e, \tag{12}$$

where $z \in \mathbb{R}^n$ is vector received by B, $y \in \mathbb{R}^n$ is vector transmitted by A, $e \in \mathbb{R}^n$ is additive noise vector at the receiver B, $A \in \mathbb{R}^{n \times n}$ is legitimate channel matrix. It is assumed that e are i.i.d. *Gaussian* vectors: $N\left(0, \sigma_e^2\right)$. $A = \left(a_{ij}\right), (i = 1 \dots n, j = 1 \dots n)$ are also matrices with i.i.d. Gaussian matrix elements: $N(0, \sigma^2)$. For Eavesdropper channel from A to E (Eve) holds the equation:

$$z' = By + e', \tag{13}$$

where $z' \in \mathbb{R}^n$ is vector received by E, $e' \in \mathbb{R}^n$ is additive noise vector at the receiver E, $B \in \mathbb{R}^{n \times n}$ eavesdropper channel matrix. It is assumed that e' are also i.i.d. vectors based on Gaussian distribution $N\left(0, \tilde{\sigma}_e^2\right)$, $B = \left(b_{ij}\right)$, $i = 1, \dots, n, j = 1, \dots, n$ with b_{ij} which are i.i.d.

Gaussian values $N(0, \sigma_w^2)$. All entries of matrices A and B are assumed to be mutually independent. Next three conditions are very important for further discussion:

- channel matrices A and B are constant for some time until user A applies encoder,
- matrix A is known exactly by legitimate users,
- matrices A and B are known exactly by eavesdropper E.

It is worth to note that the model described above is more or less valid in practice for fading channels based on *MIMO technology* if space distance between legitimate users and eavesdropper is at least several wavelengths of communication. Encoding procedure in line with [12] is the following:

$$y = Vx, \tag{14}$$

where $V \in \mathbb{R}^{n \times n}$ is orthogonal matrix taken from *singular value decomposition* (SVD) of matrix $A = USV^T$, $x \in \mathbb{R}^n$ with binary entries x_i, $i = 1, \dots, n$. Predecoding procedure in line with [12] is the following:

$$z' = U^T z = U^T Ay + e' = U^T USV^T Vx + U^T e = Sx + e', \tag{15}$$

where $U \in \mathbb{R}^{n \times n}$ is orthogonal matrix taken from SVD of matrix A, $e' = U^T e$. Since S is diagonal matrix, we get from (15) the following optimal decoding rule:

$$x' = \arg\min_{x_i} |z_i' - x_i s_i|, \quad i = 1, \dots, n, \tag{16}$$

where s_i is i-th element of diagonal matrix S, x_i are binary entries of vector x. We can see from (16) that decoding procedure has linear complexity on the number of antennas n. Eavesdropper E following to strategy of legitimate users performs the optimal decoding procedure:

$$z' = U'^T z', \tag{17}$$

where

$$z' = BVx + e' = U'S'V'^T Vx + e', \tag{18}$$

where U', V', S' are SVD of matrix B. Substituting (18) into (17), we get:

$$z'' = Cx + \tilde{e}, \tag{19}$$

where $C = S'V'^T V$, $\tilde{e} = U'^T e'$. Since matrix C is not a diagonal one in this case, their optimal decoding be the following:

$$\tilde{x} = \arg\min \|z'' - Cx\|, \tag{20}$$

where$\|\cdot\|$ is Euclidian norm in $\mathbb{R}^n$. Solution of problem (20) is known as *hard CVP problem* and it was proved in [12] that it has exponential complexity with respect to the number of antennas n if the following condition holds:

$$\sigma_w^2 \tilde{\sigma}_e^2 > n^{1/2}$$

But let us consider suboptimal decoding method after some transform, assuming that matrix C is non-singular:

$$C^{-1}z'' = x + C^{-1}\tilde{e}$$

Thus suboptimal decoding method can be implemented as follows:

$$\tilde{x}_i = \arg\min |\tilde{z}_i - x_i|, \quad i = 1, \ldots, n, \tag{21}$$

where $\tilde{z}_i$ is the i-th entry of vector $C^{-1}z''$.

We can see from (21) that complexity of suboptimal decoding procedure is linear on the number n of antennas. The efficiency of DGC can be estimated by comparing the bit error rate (BER) for optimal decoding (16) and suboptimal decoding (21). In fact if for some chosen DGC parameters the first BER is satisfactory while the second one is close to ½, then DGC can be termed as secure. In **Table 1** are presented the results of simulation for BER's by (16) and (21) denoted as p and p', respectively.

We can see from **Table 1** that for all five set of system parameters, DGC is looking as acceptable one because the case $p' \geq 0.3$ does not allow to recover a meaningful text. In fact, let us consider 32-ary symmetric noisy channel (in line with 32-ary alphabet of Russian language). Then it is easy to see that *Shannon capacity* of such channel, if every letter is encoded by five bits with BER equal to p', is:

$$\begin{aligned} C = {} & 5 + (1-p')^5 \log_2 (1-p')^5 + 5p'(1-p')^4 \log_2 \left(p'(1-p')^4\right) + \\ & +10p'^2(1-p')^3 \log_2 \left(p'^2(1-p')^3\right) + 10p'^3(1-p')^2 \log_2 \left(p'^3(1-p')^2\right) + \\ & +5p'^4(1-p') \log_2 \left(p'^4(1-p')\right) + p'^5 \log_2 p'^5 \end{aligned} \tag{22}$$

In **Table 2** are presented the values of channel capacity calculated by (22) for different values p'.

It is well known that entropy H of Russian language lies within interval 1.5÷2.5 bit/letter. This means that according to Shannon's theorem, if $H > C$, then a reading of meaningful text be impossible. In our case we get that if $p' > 0.19$ such text decoding cannot be done, that is exactly the thing for every set of parameters in **Table 1**. Moreover, we simulated transmission of Russian meaningful text over the channel

No	System parameters	n	Symbol error probability for legitimate users (p)	Symbol error probabilities for Eavesdropper (p')
1	$\sigma^2 = \sigma_w^2 = 2, \sigma_e^2 = \tilde{\sigma}_e^2 = 1$	100	0.02	0.2
2	$\sigma^2 = \sigma_w^2 = 4, \sigma_e^2 = \tilde{\sigma}_e^2 = 8$	100	0.037	0.3
3	$\sigma^2 = \sigma_w^2 = 1, \sigma_e^2 = \tilde{\sigma}_e^2 = 30$	100	0.02	0.42
4	$\sigma^2 = 2, \sigma_w^2 = 1, \sigma_e^2 = \tilde{\sigma}_e^2 = 4$	1000	5.6×10^{-3}	0.3
5	$\sigma^2 = 4, \sigma_w^2 = 8, \sigma_e^2 = \tilde{\sigma}_e^2 = 12$	1000	0.01	0.33

Table 1.
The results of simulation for BER p and p′ with different chosen parameters of DGC.

p'	0.1	0.11	0.15	0.18	0.19
C	2.65	2.5	1.95	1.6	1.5
p'	0.2	0.25	0.3	0.35	0.4
C	1.39	0.94	0.59	0.33	0.15
p'	0.41	0.42	0.43	0.44	0.45
C	0.12	0.093	0.071	0.052	0.036
p'	0.46	0.47	0.48	0.49	0.5
C	0.023	0.013	0.0058	0.0014	0

Table 2.
The values of channel capacity C for different p'.

with BER equal to 0.19 and have got that only very short words could be readable. Thus we may conclude so far, that DGC is secure against eavesdropping at least for the condition $n'_r = n_r$ and the more $n'_r < n_r$. But the following question arises—if such conclusion is true for the case $n'_r > n_r$? In **Table 3** are presented the results of simulation BER's p' for eavesdropper by suboptimal decoding rule (21) in the case of typical parameters $\sigma^2 = \sigma_w^2 = 4$, $\sigma_e^2 = \tilde{\sigma}_e^2 = 7$, $n_r = 100$ and different n'_r, where inverse matrix C^{-1} for rectangular $n_r \times n'_r$ matrix C was calculated as *Moore-Penrose pseudo-inverse matrix* to C[14].

We can see from **Table 3** that even small increasing of n'_r (on 9 antennas) compared to n_r, results in a drastic degradation of DGC because the symbol error probability p' becomes for eavesdropper very close to the probability p for legitimate users. In order to find out that such "paradox" that contradicts to our intuition appears not due to *"ill-posed" inverse matrices* [13], let us consider a theoretical proof of the bounds for the symbol correct probabilities both for legitimate users and eavesdropper. It is easy to see that decision rule (16) for legitimate users when $x_i \in (0, 1)$ is equivalent to the following relation:

$$x'_i = \begin{cases} 0 \text{ if } z''_i \le S_i/2, \\ 1 \text{ if } z''_i > S_i/2. \end{cases}$$

Thus for the symbol correct probability we get the following lower bound:

$$P\{x'_i = x_i\} \ge P\{|e_i| < S_i/2\}, \tag{23}$$

where e_i is the i-th entry of additive noise vector e in (12). Because we assumed before that $e_i \sim N(0, \sigma_e^2)$ we get from (23):

n'_r	100	101	102	103	105	107
p'	0.31	0.22	0.16	0.12	0.07	0.048
n'_r	108	109	110	120	150	
p'	0.039	0.03	0.024	0.003	7×10^{-4}	

Table 3.
Results of BER p' simulation for decision rule (21), $n_r = 100$ and different $n'_r \ge n_r$.

$$P\{x_i' = x_i\} \geq 2\Phi\left(\frac{S_i}{2\sigma_e}\right) \tag{24}$$

where $\Phi(a) = \frac{1}{\sqrt{2\pi}} \int_0^a \exp\left(-\frac{t^2}{2}\right) dt$. The decision rule (21) for eavesdropper will be equivalent to the following one:

$$\tilde{x}_i = \begin{cases} 0 \text{ if } \tilde{z}_i \leq 1/2, \\ 1 \text{ if } \tilde{z}_i > 1/2. \end{cases} \tag{25}$$

From relation (25) we get the lower bound for correct symbol probability:

$$P\{\tilde{x}_i = x_i\} \geq P\left\{ |e_i''| \leq \frac{1}{2} \right\} = 2\Phi\left(\frac{1}{2\sqrt{Var\{e_i''\}}}\right), \tag{26}$$

where e_i'' is the i-th entry of additive noise vector $e'' = C^{-1}\tilde{e}$. In order to find $Var\{e_i''\}$ let us accomplish some matrix transforms:

$$\begin{aligned} C &= BV = U'S'V'V, \\ C^{-1} &= V^T V' (S')^{-1} U'^T, \\ e'' &= C^{-1}\tilde{e} = V^T V' (S')^{-1} U'^T. \end{aligned} \tag{27}$$

Taking into account that U' is orthogonal matrix and S' is diagonal one, we get from (27):

$$Var\{e_i''\} = \tilde{\sigma}_e^2 \sum_{k=1}^{n_r} \frac{V_{ik}^2}{S_k'^2} \tag{28}$$

where V_{ik} are elements of matrix $V^T V'$ and S_k' elements of matrix S'. Substituting (28) into (27) we obtain

$$P\{\tilde{x}_i = x_i\} \geq 2\Phi\left(\frac{1}{2\tilde{\sigma}_e \sqrt{\sum_{k=1}^{n_r} \frac{V_{ik}^2}{S_k'^2}}}\right) \tag{29}$$

In order to compute theoretically the average value of symbol correct probabilities, it would be necessary to average relations (24) and (29) on the probability distribution of singular values S_k, and also on elements of channel matrices $V^T V'$. Solution to this problem requires a very crude approximations. Therefore we used simulations by (24) and (29). In **Table 4** are presented such results for both legitimate users (q) and for eavesdropper (q') with channel parameters: $\sigma^2 = \sigma_w^2 = 7$, $\sigma_e^2 = \tilde{\sigma}_e^2 = 4$, $n_t = n_r = 100$ against different numbers n_r' for eavesdropper antennas.

We can see from this Table that an increasing of the eavesdropper's antennas even till $n_r' = 105$ results in equality of values q and q' that is in line with our previous claiming. Thus we can conclude that a compromising of DGC after a small increment of the eavesdropper antenna numbers against legitimate user antenna numbers is the

n_r'	100	101	102	103	104	105
q	0.95	0.95	0.95	0.95	0.95	0.95
q'	0.71	0.75	0.8	0.87	0.9	0.95
n_r'	106	107	108	109	110	
q	0.95	0.95	0.95	0.95	0.95	
q'	0.96	0.97	0.98	0.985	0.99	

Table 4.
The symbol correct probabilities obtained by simulation of averaged bounds for q (24) and for q′ (29).

proved fact but not a consequence of an ill-conditioned matrix property. On the other hand, legitimate users executing DGC do not take for granted that the condition $n_r' \leq n_r$ holds. In the paper [14] it was proposed to change matrix V in "precoding" procedure to another matrix. In particular, authors of the current paper have proved that a choice of matrix A^{-1} as precoding one has some advantages, namely a growth of some parameter "advantage" (introduced in [14]) for legitimate users proportional to n^2 if $n = n_t = n_r = n_r'$. Such approach means that a precoding procedure is simply "canceling of channel fading".

However in the case of such precoding we face with a growing of the transmitter power. In **Table 5** are presented the results of the average transmitter power calculated for the case of precoding with matrix A^{-1}, and obtained by simulation for $n = n_t = n_r = n_r' = 100$ on different sessions with 10,000 realizations for each session.

We can see from **Table 5** that the use of *inverse precoding* results in a drastic growing of the required transmission power and to a large fluctuations on each of sessions that makes such approach impracticable (we note that such power was equal to be $n^2 = 10^4$ only for ordinary encoding by matrix V). It seems to be acceptable to use for a precoding $V = A^{-1}$ only in the case with the required transmitter power $P_0 \leq P_{tr}$ where P_{tr} is some reasonable threshold and ordinary matrix V otherwise. But such approach requires further investigations.

There is one more problem with practical implementation of DGC. It is a correct estimation of the channel matrix A by legitimate users. Of course, they may do it by a sending of special test signal from user A to user B and back from B to A during a coherent time of legitimate channel, when matrix A holds practically constant. But any way it will result in some matrix A corruption. In **Table 6** are given our simulation results for symbol error probabilities that obtain legitimate users if they estimate elements of the channel matrix A with Gaussian noise error having variance σ_ε^2. We can see from **Table 6** that a correctness of matrix element estimation affects very strong on the symbol error probabilities. It is possible to neglect such incorrectness

Session numbers	1	2	3	4	5
P_0	4.5×10^5	1.8×10^7	9.1×10^5	2.1×10^6	8.8×10^4
Session numbers	6	7	8	9	10
P_0	5.9×10^5	3.4×10^4	5.2×10^4	2.5×10^5	1.08×10^5

Table 5.
Average transmitter power P_o with precoding matrix A^{-1} for different sessions.

σ^2	σ_e^2	σ_ε^2				
		1	**0.1**	**0.01**	**0.001**	**0**
2	3	0.1677	0.0587	0.0271	0.0208	0.02
3	8	0.1332	0.053	0.0383	0.0378	0.037
50	4	0.0364	0.0126	0.0093	0.0088	0.0084

Table 6.
Results of symbol error probabilities for legitimate users under the incorrect estimation of channel elements with variance σ_ε^2.

only if signal-to-noise ratio for legitimate users is about 30-40 dB that is sufficiently high requirement.

Concluding the Section 3 we may say that theoretically would be interesting to develop further DGC in the direction of improving precoding algorithm. But according to our opinion, practical implementation of such approach is very sensitive to channel and system parameters (SNR for eavesdropper and their antenna numbers). It is worth to note also that it is not so dangerous to face with unfavorable parameters for DGC implementation as the fact that these parameters cannot be controlled by legitimate users and hence they are unable to match with parameters of DGC. In the Section 5 we consider protocol that is completely invariant to channel parameters.

4. EVSkey scheme proposed by Qin and Ding

In this Section we consider the second scenarios of keyless cryptography the purpose of which is to share secret keys between legitimate users communicating with each other over channels and next to execute the shared key for ordinary key-based cryptography. In the current section we consider key sharing protocol proposed by G. Qin and Z. Ding in the paper [11], called *EVSkey Scheme* and declared secure by their authors. This KSP is presented in **Figure 3**. Before a performance of KSP A and B generate their own random unitary matrices $X_A, X_B \in \mathbb{C}^{n\times n}$ as well as random Ginibre matrices $G_A, G_B \in \mathbb{C}^{n\times n}$, where n is the number of antennas employed by each of the users, and length of pilot signals too. Matrices H_{AB}, H_{BA} are $n \times n$ channel matrices with independent Gaussian matrix elements distributed according to $(h_{AB}), (h_{BA}) \sim CN(0,1)$. N_{A_1}, N_{B_1} are additive white Gaussian noise (AWGN) matrices $(n_{A_1})_{ij}, (n_{B_1})_{ij} \in CN(0,\sigma^2)$ of proper noises for legitimate users A and B, respectively.

Let us introduce the following matrices: $P = H_{BA}G_B, Q = H_{AB}G_A$. Then PQ and QP can be estimated by users via least square method as

$$PQ \approx Y_{A_2}(X_A)^{-1}, \quad QP \approx Y_{B_2}(X_B)^{-1}$$

It is well known [13] that square matrices of the same size PQ and QP have the same characteristic polynomials (CP):

$$CP[PQ] = CP[QP] \tag{30}$$

Thus from (30) we conclude that the legitimate users A and B are able to extract the same CP after a completion of four-steps KSP through noiseless channels although matrices PQ and QP can be different.

It was claimed in [11] that EVSkey Scheme is secure against interception of key bits by eavesdropper E. Let us show that, in fact, such KSP is insecure because eavesdropper E is able to intercept the key bits if she intercepts simultaneously the following signals:

$$\begin{aligned} \tilde{Y}_{A_1} &= H_{BE}G_BX_B, \quad \tilde{Y}_{A_2} = H_{BE}G_BH_{AB}G_AX_A, \\ \tilde{Y}_{B_1} &= H_{AE}G_AX_A, \quad \tilde{Y}_{B_2} = H_{AE}G_AH_{BA}G_BX_B, \end{aligned} \tag{31}$$

where H_{AE}, H_{BE} denote E's channel matrices and under the condition that all E's noises are equal to zero.

Statement 1. For random matrices described above, say Y, the inverse matrix $(Y)^{-1}$ and, in the general case of rectangular matrices, the Moore-Penrose pseudoinverse $(Y)_P^{-1}$ matrix [13] does exist with probability one.

Proof. Indeed let H be $n \times n$ complex random matrix with i.i.d. elements which are $CN(0,1)$. Then its joint element density is [15]:

$$f(H) = (2\pi)^{-n^2} \exp\left(-\frac{1}{2}Tr\left(HH^T\right)\right),$$

where $Tr(\cdot)$ is matrix trace. Degenerate matrices ($\det H = 0$) form $2n^2 - 2$ dimension manifold. $f(H)$, being non-singular, entails zero probability for H to hit any manifold of lower than $2n^2$ dimension. Thus, probability of $\det H \neq 0$ equals one. Such matrices stay invertible being multiplied by unitary matrices G, X. □

Statement 2. Let $EV(Y)$ denote the set of matrix Y eigenvalues. Then

$$EV(Y) = EV(PQ) = EV(QP),$$

where

$$Y = \tilde{Y}_{A_2}\left(\tilde{Y}_{B_1}\right)^{-1}\tilde{Y}_{B_2}\left(\tilde{Y}_{A_1}\right)^{-1}. \tag{32}$$

Proof. Substituting (31) into (32), we get:

$$Y = H_{BE}G_BH_{AB}G_AH_{BA}G_B(H_{BE}G_B)^{-1} = (H_{BE}G_B)QP(H_{BE}G_B)^{-1}$$

The last relation means that matrix Y is *similar* to matrix QP and thus $EV(Y) = EV(QP)$ for any matrices H_{AE}, H_{BE}. □

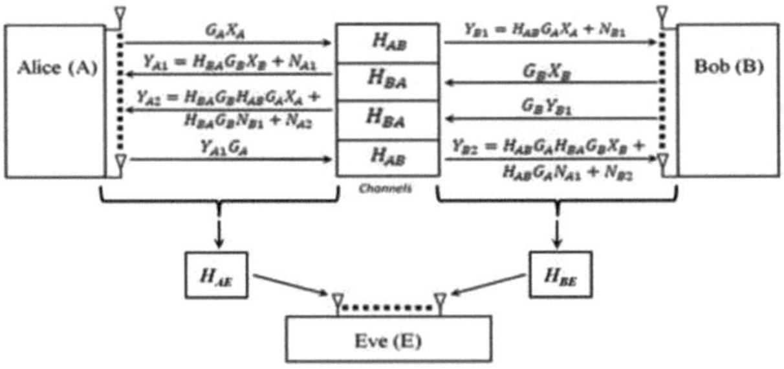

Figure 3.
The KSP corresponding to EVSkey scheme.

The statement 2 has been proved for the case of noiseless E's channels. But our simulation shows that even for noisy channels, if Eva be following to algorithm (32), she extracts the key bits with BER very close to those that have legitimate users. This fact proves finally that EVSkey Scheme is in fact *insecure and hence cannot be recommended for practical application*. However the general idea to use KSP with matrix exchange over the channels and extraction of key bits after a quantization of eigenvalues is possible if protocol be elaborated in the desired directions. Such KSP is presented in the next Section.

5. Novel key sharing protocol for public constant noiseless channels and without any cryptographic assumptions

Let us introduce novel KSP designed for its use in constant, public and noiseless communication channels. This protocol has the following distinctions in comparison with other KSP's:

- it executes over public, constant parameter, noiseless channels (like internet),
- no cryptographic assumptions are needed in order to provide its security and quantum computers cannot break it,
- this KSP provides tradeoff between key reliability and security for given size of key strings,
- security of protocol is estimated in terms of Shannon information leaking to eavesdropper,
- a good key bits statistics can be provided due to their generation by hardware devices,
- the size of traffic for execution of protocol is acceptable for ordinary users,
- the number of protocol steps is minimized and equal to 2.

In **Figure 4** it is presented KSP protocol for execution in public constant noiseless channels that we call *KSP-PCN*. Here G and P are $n \times n$ random matrices generating A and B in advance as well as $n \times n$ matrices of artificial noises N_{A_1}, N_{B_1}. We believe that all matrix elements are independent of each and Gaussian distributed.

At first step, these matrices are transmitted over constant public noiseless channel to opposite party. Next, each of users A and B calculates $\tilde{K}_A$ and $\tilde{K}_B$ that are noisy version of matrices GP, PG, respectively, and extract eigenvalues from these matrices. After a quantization procedure of eigenvalues A and B obtain "raw bits" of the shared key $\tilde{\tilde{K}}_A$ and $\tilde{\tilde{K}}_B$, respectively. Eavesdropper E intercepts signals $G + N_A$, $G + N_B$. It extracts eigenvalues from the matrix product $(G + N_A)(G + N_B)$ and subsequently quantizes them to get eavesdropper's "raw key" K_E. Simulation of KSP shows that the of key bit error probabilities for legitimate users P_l occur very close to that ones P_e of eavesdropper. Hence we need such additional protocol that be able to "diverse" these probabilities. Such protocol that was called *Preference Improvement of the Main Channel* (PIMC) works as follows: user A (or B) repeats S times each "raw bit", while opposite

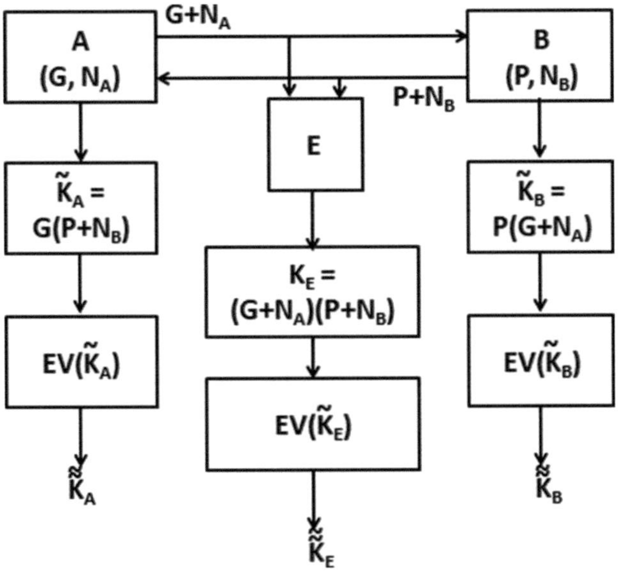

Figure 4.
Two-step KSP-PCN protocol.

user accepts such S-blocks if, and only if, they consist of the same S bits (zeros or ones),otherwise opposite user inform initiator of protocol about inadequate receiving blocks. Supposing that bit errors are independent, one from another, PIMC protocol provides the following BER between A and B:

$$\tilde{P}_l = \frac{P_l^S}{(1-P_l)^S + P_l^S} \tag{33}$$

At the same time eavesdropper E also intercepts S-blocks with BER P_e and controls public channel over that was transmitting information regarding acceptance or rejection of S-blocks between legitimate users. E takes decisions about each S-block using *majority rule*. If E's channel is believed statistically independent on the legitimate channel, then the BER after such decision will be (for odd number S):

$$\tilde{P}_e = \sum_{i=\frac{S+1}{2}} \binom{S}{i} P_e^i (1-P_e)^{S-i} \tag{34}$$

In order to provide a repetition of bit's blocks and to improve key bit statistics, it was proposed to use the scheme of key bit generation shown in **Figure 5**.

We can see from **Figure 5** that user B generates *truly random binary string* γ that is XOR-ed with B's raw bit string $\tilde{\tilde{K}}_B$ and the sum is transmitted over public noiseless channel to user A who adds this string with raw bits string $\tilde{\tilde{K}}_A$ in order to get:

$$K_A = \tilde{\tilde{K}}_B \oplus \gamma \oplus \tilde{\tilde{K}}_A = \tilde{\tilde{K}}_A \oplus \varepsilon_{AB} \oplus \tilde{\tilde{K}}_A \oplus \gamma = \gamma \oplus \varepsilon_{AB}, \tag{35}$$

where ε_{AB} is noise string between raw strings $\tilde{\tilde{K}}_A$ and $\tilde{\tilde{K}}_B$, "$\oplus$" is operation of bitwise modulo two addition. In such version only one user has to repeat S times each

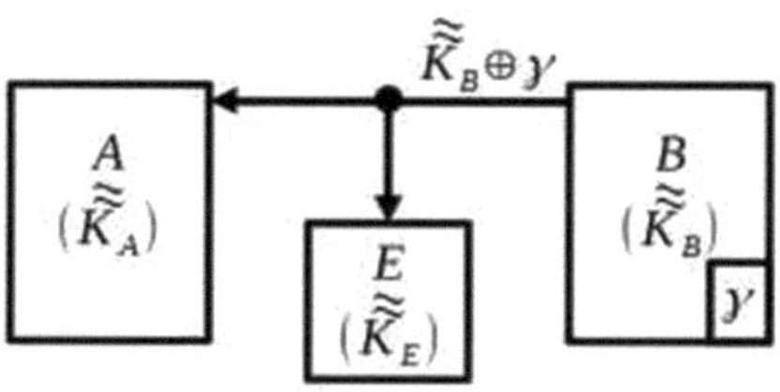

Figure 5.
Additional key-transformed protocol.

bit of γ in order to perform PIMC protocol. From now on a string γ will be considered as a final key bit string between A and B. At the same time E receives $\tilde{\tilde{K}}_B \oplus \gamma$ over public noiseless channel and possessing her string $\tilde{\tilde{K}}_E$ be able to add her intercepted bit string K_E as follows:

$$K_E = \tilde{\tilde{K}}_E \oplus \gamma \oplus \tilde{\tilde{K}}_B = \tilde{\tilde{K}}_B \oplus \varepsilon_{BE} \oplus \gamma \oplus \tilde{\tilde{K}}_B = \gamma \oplus \varepsilon_{BE}, \tag{36}$$

where ε_{BE} is noise string between raw string $\tilde{\tilde{K}}_E$ and $\tilde{\tilde{K}}_B$. In order to optimize KSP with point of reliability, security and key string size, it is necessary to select the following parameters:

- sizes of matrices n,
- number of bit repetitions S,
- variances of additive artificial noises σ^2.

It is worth to note that in the proposed KSP (see **Figure 4**), unlike the earlier presented protocols (see [5]), all parameters are *under the control of legitimate users*. This means that we can take for granted reliability and security of the shared key string as well as its size regardless of properties of eavesdropper's channel.

Although the formulas (33) and (34) of BER for both legitimate users and eavesdropper have been already proved they have to be specified by simulation because our assumption regarding statistical independence of errors is valid only partly. In **Tables 7–9** are presented the results of simulation for BER's $\tilde{P}_l$ and $\tilde{P}_e$ for given parameters σ^2, the matrix sizes n and parameter S. (They were chosen, of course, from a huge simulation results which have been removed simply as unsuitable.)

We can see from these Tables that a choice of matrix size $n = 1$ (i.e. replacing matrices with integer numbers) is inadmissible because the probabilities $\tilde{P}_l$ and $\tilde{P}_e$ occur close to one another for all parameters S and σ^2. The choice of matrix size $n = 4$ also cannot be recommended because it is inferior to the case with $n = 64$. In the last case we can see the most large difference of the BER's $\tilde{P}_l$ and $\tilde{P}_e$ for some parameters S and σ^2. But unfortunately the BER $\tilde{P}_l$ and $\tilde{P}_e$ cannot be chosen as final results.

In fact, the probability $\tilde{P}_l$ has to be decreased in order to provide an acceptable reliability of the shared key with the key length at least 256 bits. The probability $\tilde{P}_e$ is quite unacceptable because it results in a large leakage of the key information to the eavesdropper E.

Therefore, it is necessary first of all to correct errors in legitimate channel, suppose, using error correcting codes and next to amplify security of the shared bit string

s	σ^2				
	0.1	**0.2**	**0.3**	**0.4**	
1	0.296	0.343	0.37	0.387	$\tilde{P}_l$
	0.222	0.269	0.299	0.316	$\tilde{P}_e$
3	0.0523	0.0703	0.0974	0.113	$\tilde{P}_l$
	0.0407	0.0643	0.0863	0.103	$\tilde{P}_e$
5	0.00727	0.0122	0.0174	0.0224	$\tilde{P}_l$
	0.00638	0.0125	0.0189	0.0268	$\tilde{P}_e$
7	0.00124	0.00209	0.0039	0.00673	$\tilde{P}_l$
	0.00133	0.00332	0.00571	0.0119	$\tilde{P}_e$

Table 7.
The results of simulation for BER $\tilde{P}_l$ and $\tilde{P}_e$ for matrix size $n = 1$ (integers instead of matrices), given different noise variances σ^2 and the number of repetitions S.

against eavesdropping. Let us apply so called enhance of privacy amplification procedure described in [16]. We recall that *privacy amplification* procedure can be performed in two stages: firstly with the use of hashing by hash function taken randomly from universal$_2$ class, and secondly, by special "puncturing" of the hashed string [16]. Then the upper bound for Shannon's information I leaking to eavesdropper be given by the following [16]:

$$I \leq \frac{2^{-(k-t_c-l_0-r)}}{\alpha \ln 2}, \tag{37}$$

where k is the string, generated by A and B after a completion of PIMC protocol, t_c is the Renyi (or collision) information obtained by E via eavesdropper BSC channel with BER $\tilde{P}_e$, r is the number of check bits sent by one of legitimate users to another one in order to reconcile their key strings finally, l_0 is the length of the final key string, α is a coefficient that approaches to 0.42 for any fixed r as k and $k - r$ are increasing. It has been proved in [17] that in order to satisfy Shannon's theorem about reliable communication over noisy channel and provide an exponential decreasing of information leaking to eavesdropper E, minimum sum $k_0 + r_0$ is provided with:

$$r_0 = H_L \frac{\lambda + l_0}{CH_C - H_L}, \quad k_0 = C\frac{\lambda + l_0}{CH_C - H_L}, \tag{38}$$

where

$$\begin{gathered}
H_L = 1 - C = -\tilde{P}_l \log_2 \tilde{P}_l - (1 - \tilde{P}_l) \log_2 (1 - \tilde{P}_l), \\
H_C = -\log_2 \left(\tilde{P}_e^2 + (1 - \tilde{P}_e)^2\right), \\
C = 1 + \tilde{P}_l \log_2 \tilde{P}_l + (1 - \tilde{P}_l) \log_2 (1 - \tilde{P}_l), \\
\lambda = k - t_c - l_0 - r, \\
t_c = k - kH_C.
\end{gathered}$$

s	σ^2				
	0.1	**0.2**	**0.3**	**0.4**	
1	0.26	0.294	0.309	0.317	$\tilde{P}_l$
	0.212	0.252	0.271	0.279	$\tilde{P}_e$
3	0.031	0.0428	0.0477	0.051	$\tilde{P}_l$
	0.0333	0.0427	0.0562	0.0604	$\tilde{P}_e$
5	0.00293	0.00443	0.0048	0.00535	$\tilde{P}_l$
	0.00528	0.00822	0.0106	0.0118	$\tilde{P}_e$
7	0.000182	0.000464	0.00061	0.000644	$\tilde{P}_l$
	0.00108	0.000195	0.000308	0.00358	$\tilde{P}_e$

Table 8.
The results of simulation for BER $\tilde{P}_l$ and $\tilde{P}_e$ for matrix size $n = 4$ given different noise variances σ^2 and the number of repetitions S.

s	σ^2				
	0.1	**0.2**	**0.3**	**0.4**	
1	0.0816	0.0851	0.0859	0.0963	$\tilde{P}_l$
	0.0922	0.117	0.140	0.154	$\tilde{P}_e$
3	0.00362	0.00402	0.00407	0.00396	$\tilde{P}_l$
	0.0185	0.0439	0.0673	0.0822	$\tilde{P}_e$
5	0.000193	0.000294	0.000258	0.000226	$\tilde{P}_l$
	0.00875	0.0314	0.053	0.0699	$\tilde{P}_e$
7	$1.44 \cdot 10^{-5}$	$3 \cdot 10^{-5}$	$1.79 \cdot 10^{-5}$	$1.68 \cdot 10^{-5}$	$\tilde{P}_l$
	0.00529	0.0256	0.0455	0.059	$\tilde{P}_e$

Table 9.
The results of simulation for BER $\tilde{P}_l$ and $\tilde{P}_e$ for matrix size $n = 64$ given different noise variances σ^2 and the number of repetitions S.

Let us adopt the following parameters from **Table 9**: $n = 64, S = 5, \sigma^2 = 0.4, \tilde{P}_l = 0.00022, \tilde{P}_e = 0.0699$. After a simulation of LDPC decoding procedure in line with [18] we select LDPC code with parameters $k = 24039, r = 961$, that provides for the final key length $l_0 = 3659$ the block error probability after decoding $P_{ed} \approx 2.5 \cdot 10^{-3}$ and information leakage $I_0 \approx 9 \cdot 10^{-4}$ bits. If we select LDPC code with parameters $k = 50001, r = 1999$, we get for final key length $l_0 = 7624$ the error probability after error correction $P_{ed} = 8 \cdot 10^{-4}$ and information leakage $I = 1.4 \cdot 10^{-3}$ bits. In the paper [17] has been proved the formula for amount of the traffic bytes that is needed in order to perform KSP described above:

$$Traffic = Tr = 2.66 \cdot 10^{-6} \frac{Skn}{\left((1 - P_l)^S + P_l^S\right)^2} \text{MB} \quad (39)$$

Figure 6.
View of random number generator Crypton USB-DRN.

Substituting the corresponding parameters chosen for KSP, we get that $Tr \approx 32$ MB that is acceptable for practical implementation.

As it can be seen from a description of proposed KSP, the generators of random numbers are needed for its implementation. Moreover it is impossible to use standard program-oriented generator (like MT19937) because it can be vulnerable to sequence prediction attack. Thus it is necessary to use *hardware generator of random numbers*. We propose to take such generator as Crypton USB-DRN that is manufactured by company "Ankad" [19]. Photo of this device is shown in **Figure 6**. We tested this device on standard NIST tests and concluded that it passes all of them except two last ones from the list of 15-ary NIST tests. Experiment showed that it is required at most about 20 minutes in order to generate key string according to this KSP.

For our KSP is required (as for any such protocol) to perform authentication procedure—otherwise the adversary could impersonate legitimate users and eventually share with them a common key. It is possible to use different authentication methods: short key, the Needham-Schroder protocol [20] or pairing procedure during the face to face device meeting [21].

6. Conclusion

In the current chapter we have presented four different systems related with keyless cryptography. The first two of them consider such systems that do not require any key distribution in advance. The first one is based on protocol with feedback that uses public key cryptosystems but it does not require any key distribution in advance, even public one. It executes commutative cryptography but, unfortunately, it is a poor choice. It would be very interesting to find at least one of symmetric strong block cipher or post-quantum public cryptosystems that belong to commutative ones (or may be to prove that such cryptosystem does not exist at all). The second example in our "gallery" of keyless cryptography was Dean-Goldsmith one. It is based on physical layer properties and has unquestionable theoretical interest. But unfortunately it is impractical because requires from eavesdroppers unrealistic conditions. The third example is related with key sharing without some short subkeys sharing in advance. Unfortunately authors' claiming that such system is secure, occurs wrong. But their scheme can be significantly modified (see version four) to be in fact secure. According to our opinion, the fourth key sharing protocol is the first attempt to distribute keys by secret manner executing over such very popular channel as internet (or over any other public and noiseless channel). It provides easy way for a confidential communication between ordinary internet users. In the future it will be interesting to investigate exchange by integers (but not matrices) that results in, for sure, to a decreasing of channel traffic. Elaboration of reliable authentication system is also interesting both for theory and practice.

Author details

Valery Korzhik[1]*, Vladimir Starostin[1], Victor Yakovlev[1], Muaed Kabardov[1], Andrej Krasov[1] and Sergey Adadurov[2]

1 The Bonch-Bruevich State University of Telecommunications, Saint-Petersburg, Russia

2 JSC VNIIZHT, Moscow, Russia

*Address all correspondence to: val-korzhik@yandex.ru

References

[1] Alpern B, Schneider FB. Key exchange using 'keyless cryptography. Information Processing Letters. 1983;**16**(2):79-81. DOI: 10.1016/0020-0190(83)90029-7

[2] Korzhik V. Keyless cryptography. In: The 9-th International Conference on System Administration, Networking and Security; Orlando, USA. 2000

[3] Korzhik V, Bakin M. Information-theoretically secure keyless authentication. In: Proceedings of the 2000 IEEE International Symposium on Information Theory; 2000 June 25-30; Sorrento, Italy. Piscataway, NJ (USA): IEEE Press; 2000. Available from: IEEE Xplore. DOI: 10.1109/ISIT.2000.2

[4] Korzhik V. Keyless Cryptography. Invited Talk at Security Seminar at CERIAS Purdue University. 2001. Available from: http://www.cerias.purdue.edu/secsem/abstracts.html#Fall

[5] Mukherjee A, Fakoorian SAA, Huang J, Swindlehurst AL. Principles of physical layer security in multiuser wireless networks: A survey. IEEE Communications Surveys & Tutorials. 2014;**16**(3):1550-1573. DOI: 10.1109/SURV.2014.012314.00178

[6] Schneier B. Applied Cryptography. Montreal: JW Publ. Inc.; 1994

[7] Menezes AJ, van Oorschot PC, Vanstone SA. Handbook of Applied Cryptography. Boca Raton: CRC Press; 1997. DOI: 10.1201/9780429466335

[8] Shor PW. Polynomial-time algorithms for prime factorization and discrete logarithms on a quantum computer. In: Proceedings of the 35th Annual Symposium on Foundations of Computer Science; 1994 November 20–22; Santa Fe, USA. Los Alamitos: IEEE Press; 1994. pp. 124-134. DOI: 10.1109/SFCS.1994

[9] Dyakonov MI. Is fault-tolerant quantum computation really possible? In: Luryi S, Xu J, Zaslavsky A, editors. Future Trends in Microelectronics. Hoboken, NJ, USA: John Wiley & Sons; 2007. pp. 4-18

[10] Bennett CH, Bessette F, Brassard G, Salvail L, Smolin J. Experimental quantum cryptography. Journal of Cryptology. 1992;**5**:3-28

[11] Qin D, Ding Z. Exploiting multi-antenna non-reciprocal channels for share secret key generation. IEEE Transactions on Information Forensics and Security. 2016;**11**(10):2691-2705

[12] Dean T, Goldsmith AJ. Physical-layer cryptography through massive MIMO. In: Proceedings of the 2013 IEEE Information Theory Workshop; 2013 September 9-13; Seville, Spain. Piscataway, NJ, USA: IEEE; 2013. pp. 1-5

[13] Ben-Israel A et al. Generalized Inverses: Theory and Applications. NY: Springer; 2003

[14] Steinfeld R, Sakzad A. On massive MIMO physical layer cryptosystem. In: Proceedings of the 2015 IEEE Information Theory Workshop; 2015 October 11-15; Lotte City Hotel Jeju, Jeju Island, Korea. Piscataway, NJ, USA: IEEE; 2015. pp. 292-296. Available from: http://ieeexplore.ieee.org/servlet/opac?punumber=7355548.

[15] Edelman A. Eigenvalues and condition numbers of random matrices PhD thesis: Cambridge, Massachusetts, USA: Massachusetts Institute of Technology; 1989

[16] Korzhik V, Morales-Luna G, Balakirsky V. Privacy amplification theorem for noisy main channel. Lecture Notes in Computer Science. 2001;**2200**: 18-26

[17] Korzhik V, Starostin V, Kabardov M, Gerasimovich A, Yakovlev V, Zhuvikin A. Protocol of key distribution over public noiseless channels executing without cryptographic assumption. International Journal of Computer Science and Application. 2020;**17**(01): 1-14

[18] Fossorier MPC, Mihaljevic M, Imai H. Reduced complexity iterative decoding of low density parity check codes based on belief propagation. IEEE Transactions on Communications. 1999; **47**(5):673-680

[19] Podorozhnyi IV. KBDZ.469435.052 RE. Obzor apparatnykh generatorov sluchainykh chisel//Molodoy uchenyi (in Russian). Available from: http://moluch.ru/archive/105/24688/ [Accessed: June 24, 2020]

[20] Needham RM, Schroeder MD. Using encryption for authentication in large network of computers. Communications of the ACM. 1978;**21**:993-999

[21] Jin R et al. MagPairing: Pairing smartphones in close proximity using magnetometer. IEEE Transactions on Information Forensics and Security. 2016;**6**:1304-1319

Chapter 7

Perspective Chapter: Distinguishing Encrypted from Non-Encrypted Data

Eric Järpe and Quentin Gouchet

Abstract

Discriminating between encrypted and non-encrypted information is desired for many purposes. Much of the efforts in this direction in the literature is focused on deploying machine learning methods for the discrimination in streamed data which is transmitted in packets in communication networks. Here, however, the focus and the methods are different. The retrieval of data from computer hard drives that have been seized from police busts against suspected criminals is sometimes not straightforward. Typically the incriminating code, which may be important evidence in subsequent trials, is encrypted and quick deleted. The cryptanalysis of what can be recovered from such hard drives is then subject to time-consuming brute forcing and password guessing. To this end methods for accurate classification of what is encrypted code and what is not is of the essence. Here a procedure for discriminating encrypted code from non-encrypted is derived. Two methods to detect where encrypted data is located in a hard disk drive are detailed using passive change-point detection. Measures of performance of such methods are discussed and a new property for evaluation is suggested. The methods are then evaluated and discussed according to the performance measures.

Keywords: likelihood ratio, change-point detection, cryptology, compression, uniform distribution

1. Introduction

The discrimination of encrypted data from other kinds of data is of interest in many areas of application. For instance for making other applications work for the communication traffic in a network where the means for application may depend on whether the traffic data is plaintext/cleartext, compressed, encrypted or encoded in some way. Also, there may be security reasons (e.g. the uncontrolled flow of encrypted data of which some may be transmitted for malicious purposes) which could be an argument for better network supervision tools. To these ends, various methods, mainly of machine learning, have been suggested through the last decades.

1.1 Methods for discrimination in case of streamed data

As a foundation for the treatment, some kind of preprocessing is due. Among methods for this step are the chi-square statistic, Shannon entropy, Kolmogorov-Smirnov statistic, Lilliefors test, Cramér-von Mises statistic, discrete runs test, autocorrelation's test and the index of coincidence test to mention some [1, 2].

In [1] the problem of filtering encrypted data upon traffic monitoring which is outbound from a computer or a network, so-called egress filtering, is considered. Many aspects and properties of the problem are brought to attention and an extensive account for various evaluation techniques is mentioned.

Different measures for discriminating a given distribution from the normal distribution are the topic in [2]. For discrimination between encrypted and non-encrypted data other distributions than the normal are more relevant but some of the measures considered in this paper can also be used for the more general case.

In [3] machine learning methods are deployed to a fully automated method to the distinction between encrypted and compressed data, claimed to be the first of its kind. A dataset for further evaluation and benchmarking is also provided.

A method for discriminating between compressed and encrypted data in the case when the data is voice over IP with constant and variable bit rate codecs is proposed in [4]. The method is evaluated utilizing the NIST [5] and the ENT [6] test suits.

Reference [7] does not suggest a solution to the discrimination problem but rather a method for fast and distribution true simulation of data reflecting the properties of encrypted and compressed data respectively.

In Ref. [8] a classification procedure, based on Gaussian mixtures and hidden Markov models, is detailed. An elaborate comparative evaluation is carried out taking 9 other attempts to solve the same problem into account. The study concludes that the proposed method renders a better classification at a lower computation complexity.

A lucid convolution neural network method to classify data into being encrypted or non-encrypted is presented by [9]. The proposed method is thoroughly evaluated for various kinds of network traffic and some different performance measures in the field of machine learning.

1.2 Methods for discrimination in case of static data

A different need for discrimination between encrypted and unencrypted data is the following. In police investigations concerning heavy criminality and organized crime where evidence of criminal activity is residing as data on a computer hard disk drive (HDD) the capability to distinguish encrypted files from non-encrypted ones may be primordial [10–12]. When the police seize an HDD containing data belonging to a suspected criminal, that data can be material of evidence in a subsequent trial. But criminals often try to make sure that the police will be able to use that data. Possible actions to obstruct data access are then to encrypt and/or to *quick delete* that data. In case of quick delete of data, the pointers to the files are destroyed but the content of the files is still left. The other action is to encrypt files. Using state of the art tools for encryption of files (such as VeraCrypt, Bitlocker, Ciphershed and thereby provided ciphers) there is no general procedure of breaking the encryption other than brute force attack by guessing the cipher algorithm and systematically guessing the encryption key [13]. If some of the files are encrypted and others not, it is then a delicate matter to distinguish between the two if the code has been quick deleted. Nevertheless, the importance of discriminating between encrypted code and plaintext is also stressed by the fact that brute force attacking all clusters of data on the hard drive would be an impossible task while being able to separate the encrypted code from the non-encrypted would make a substantial improvement of the chances for successful cryptanalysis. The police authorities usually have software to brute force those encrypted data but this procedure may be very time consuming if the amount of data is so large that different parts of it have been encrypted with different keys. In juridical cases, time is typically of the essence since the chances of success in prosecution and proceedings against a criminal is depending on deadlines (such as time of arrest, time of trial etc) any time savings in the procedure of extracting evidence from the HDDs is essential. Thus time has to be spent on the appropriate tasks: code-breaking only on the encrypted data rather than trying to decypher non-encrypted data. Given a single file, the task of determining whether it is encrypted or not is usually easy; but given a whole HDD without knowing where the encrypted files are, this is trickier.

There are several software solutions for indicating whether a file is encrypted or not, mostly checking the header of the file and looking for some known pattern like the EFS (Encryption Files System on Windows), BestCrypt, or other softwares' headers. Such alternatives cannot be used in the case when the user has performed a (quick) delete of the HDD because then the pointers to all files are lost. Nevertheless, the files remain on the HDD: upon deleting a file on an HDD, the pointers to the beginning and end of the physical space on the HDD containing the information of the file are removed. Still, the physical space on the HDD containing the information of the file is not overwritten because that operation would be slow, i.e. as slow as writing the same file again. The operating system's designers rather leave the file in the HDD intact but indicate that this location is free to host some other data. If nothing has been overwritten, this means that the file is still stored in the HDD for some time which allows recovery software to recover those files.

Recovery software might help to simply recover the files but might also overwrite the data contained in HDD which could result in loss of evidence in case of an investigation. In this case, the police would have to locate the encrypted files without using recovery software or "header-detector" software.

Here methods to locate quick deleted encrypted data are presented and detailed. First, a description of how encrypted data is different from other data is presented. This is followed by the introduction to statistical change-point methods for discrimination between encrypted and non-encrypted data. Finally, the results of these procedures are presented along with some experimental values to evaluate the methods.

However, such a method will only work on mechanical HDDs and not with flash memory devices: in flash memory (like USB memory sticks or Solid State Drive (SSD)), as soon as a file is removed it is erased from the memory because data cannot be overwritten. Therefore as soon as data is deleted, the operating system will choose to delete the pointers and the data to gain time for the time when the user will decide to save data at this location. But erasing in flash memory also takes longer since the pointer have to be deleted and all the files removed which is as long as copying new files on the device.

This application differs from the one mentioned in Subsection 1.1. In the previous case when data is commonly considered being transmitted in packets in a network while here, there is static access to the data. In the previous situation, data consists of sometimes small files where statistical methods for making a foundation indicating if the data is encrypted or not becomes weaker [14] as opposed to the situation here where there is a large amount of data possibly of both kinds, encrypted and non-encrypted. In the previous case with dynamic data in all senses—variable in size, access, time, kind—methods of machine learning which could pick up on the current circumstances were preferable while here, in the case of fixed data, fast and efficient methods of change-point detection becomes a far more advantageous choice. The machine learning methods need training data while the change-point methods can start from scratch with pre-defined parameter values optimized for the situation or with very little run-in data for calibrating the levels. The performance of the machine learning methods is still a matter of research since these are very highly structured procedures, sometimes black box techniques impossible to fully evaluate all properties of, while the change-point detection methods are long since optimized and very thoroughly evaluated from all kinds of aspects of performance through a much longer time.

2. Method

2.1 Description

If encrypted data is not uniformly distributed, the cryptosystem used to cipher those data has a bias and can therefore be attacked. For this reason, characters of the cyphertext produced by any modern high-quality cryptosystem are uniformly distributed [15, 16] i.e. the values of the bytes of the cyphertext are uniformly distributed on some character interval. Indeed, a cryptosystem that does not have this property would be weak since it would be possible to attack it on this bias (distinguishing attacks such as on the RC4 encryption algorithm). The other types of files do not possess this feature although the contents of some types of files are close to being uniformly distributed. The files coming closest without being encrypted are compressed/zipped files: those files are indeed very close to cipher files in terms of the distribution of their character's byte numbers. Albeit small there is a difference in distribution making it possible to tell compressed files and encrypted files apart.

A technique to quantify this distribution difference is by using chi-square statistic and more or less performing a chi-square test (see [17]) to tell whether the data is suspiciously uniform or not. Another classical method is to calculate the Kolmogorov-Smirnov distance, see e.g. [1, 2] for a more extensive account of ways to indicate whether data are encrypted or not. Procedures building on such preprocessors can then be defined.

One attempt to exploit this distribution difference utilizing the chi-square statistic is [18] where a method to automate the discrimination between encrypted and non-encrypted (i.e. most critically compressed) data into a method with impressing performance. Another approach could be to use means of anomaly detection in the theory of machine learning to develop an adaptive method. The proposed method, however, stems from using statistical change-point detection. The anticipated advantage with this could be the mathematically proven optimality with these methods in terms of efficiency and accuracy under the given assumptions. For many applications, the assumptions of entirely relying on the distribution of the data, in-control and out-of-control are commonly a subject of great controversy. Here, the situation is different though, since the hard drive and its data is not a dynamic system where the content changes its distribution owing to outer time-depending circumstances.

Before going into detail about these methods the preprocessing techniques are introduced.

2.2 Distribution of encrypted data

The working hypothesis is that data (i.e. characters) constituting encrypted files are uniformly distributed, while data of non-encrypted files are not (i.e. differently distributed depending on which type of non-encrypted files). The goal now is to be able to tell an encrypted file from a non-encrypted one.

Let us assume the data constitutes of characters divided into clusters, $c_1, c_2, c_3, \ldots$ with N characters in each cluster. The characters that are used range over some alphabet of a set of possible forms. Merging these forms into K classes, the counts O_{kt} of occurrences of class k characters in cluster t are observed. One method of measuring distribution agreement is by means of a χ^2 test statistic, $Q_t = \sum_{k=1}^{K}(O_{kt} - E_{kt})^2/E_{kt}$ where E_{kt} are the expected counts of occurrences of characters in class k, cluster t. Under the hypothesis of uniformly distributed characters, the expected counts of occurrences within each class is $E_{kt} = \frac{N}{K}$ reducing the statistic simplifies to

$$Q_t = \frac{K}{N}\sum_{k=1}^{K} O_{kt} - N$$

the values of which are henceforth referred to as Q scores. Large Q scores indicate deviance from the corresponding expected frequencies E_{kt}. The smallest possible Q score being 0 would be attained if all $E_{kt} = O_{kt}$. The expected value, E_{kt}, in each class, should not be smaller than 5 for the statistic to be relevant (5 is a commonly used value; other values like 2 or 10 are sometimes used depending on how stable the test statistic should be to small deviances in tail probabilities). Therefore one should use at least 5 kB of data in each cluster to enable this test to be relevant. But the larger the number of bytes that are in each cluster, the worse the precision to detect encrypted file values: indeed, if too many bytes were detected as being unencrypted (but were actually encrypted) a large amount of encrypted data might not be detected in the procedure.

Here, the alphabet used was the numbers $0, 1, \ldots, 255$ representing the possible size in bytes of the characters in the data. These numbers were divided into $K = 8$ classes (class 1: values of bytes in $\{0, 1, 2, \ldots, 31\}$ to class 8: values of bytes in $\{224, 225, 226, \ldots, 255\}$) and the clusters of size $N = 64$ bytes making the expected values in each class $E_{kt} = 8$. Assuming that encrypted data are uniformly distributed, the Q scores based on counts of characters in encrypted data are χ^2 distributed, see **Figure 1** and since 8 classes were chosen, the number of degrees of freedom is $8 - 1 = 7$.

2.3 Distribution of non-encrypted data

For non-encrypted data, the distribution is more complicated. Each type of file has its distribution. Consequently, the standardized squared deviances from expected counts under the assumption about uniform distribution are larger and so are the Q scores of the χ^2 statistic. However, two problems emerge. Firstly, the size of these increased deviances depends on the type of data—i.e. is the data a text file, an image, a compiled program, a compressed file or some other kind of file?—and how should this information be properly taken into account? Secondly, what is the distribution of the Q score in the case when the data are not encrypted?

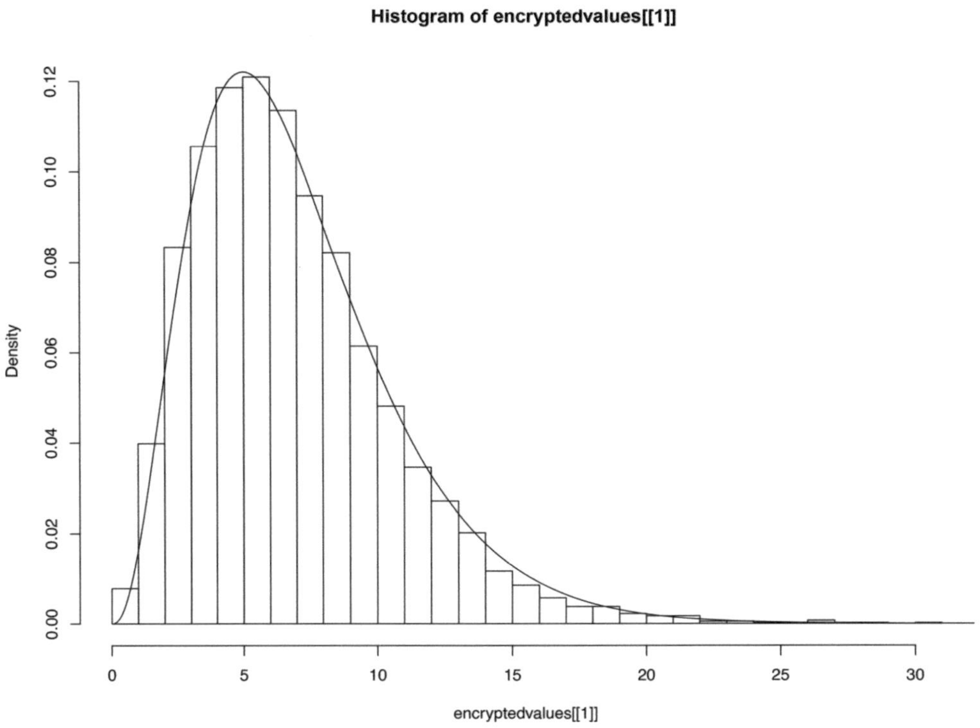

Figure 1.
Distribution of the Q scores of encrypted files (obtained by using more than 5000 files) with the distribution function of $Q \in \chi^2(7)$.

To develop a method for distinguishing between encrypted and non-encrypted data it is sufficient to focus on the non-encrypted that is most similar to the encrypted and this turns out to be compressed data. Other types of files such as images, compiled programs etc. commonly render higher Q scores and are therefore indirectly distinguished from encrypted data by a method that is calibrated for discriminating between encrypted and compressed data. About the second question, this is not readily answered. Rather, we just suggest modeling the Q score as being scaled χ^2 distributed, i.e. the Q score is assumed to have the distribution of the random variable αX where $\alpha > 1$ and $X \in \chi^2$. The validity of this approach is sustained by empirical evaluation based on more than 5000 compressed files. The resulting empirical distribution of their Q scores and the distribution of αX where X is χ^2 distributed and the value of $\alpha = 1.7374$ was estimated by the least square method was plotted in **Figure 2**.

2.4 Change-point detection

Change-point detection [19–23] is a field of mathematical statistics where the object is to quickly and accurately detect a shift in distribution from on-line observation of a random process. This can be done actively (stop collecting the data as soon as a shift is detected) or passively (continue collecting the data even if a shift is detected in order to detect more shifts). Here passive on-line change-point detection was used to detect if the data from an HDD shifts from non-encrypted to encrypted and vice versa. The change-point detection method is a stopping rule

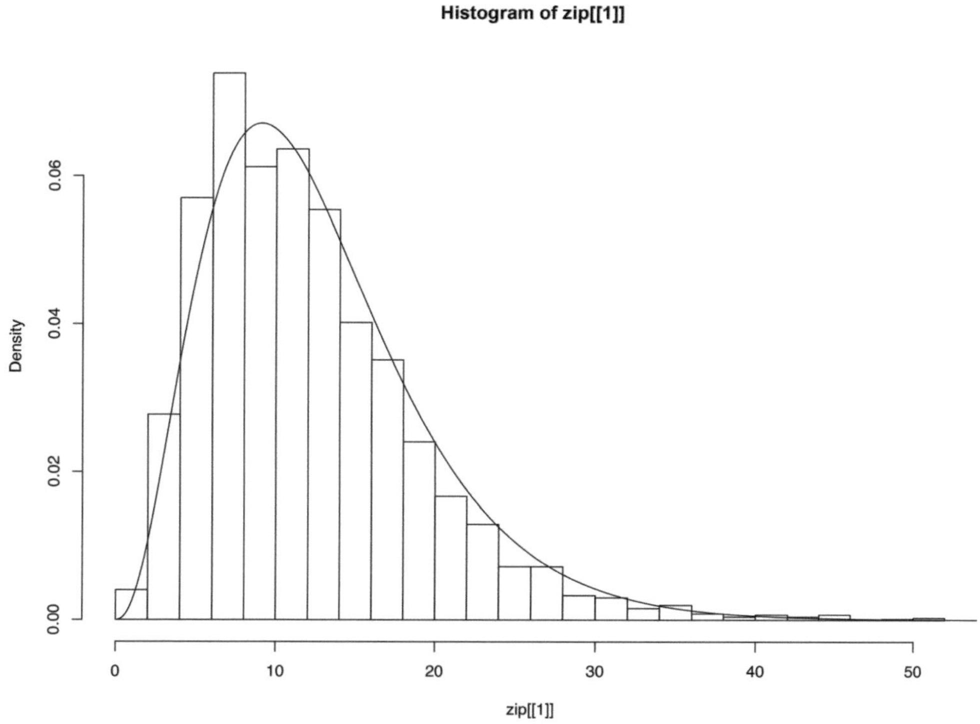

Figure 2.
Distribution of the Q scores of compressed files (obtained by using more than 5000 files) with the distribution function of $Q = 1.7374 \cdot X$ *where* $X \in \chi^2(7)$.

$$\tau = \inf \{t > 0 : a_t > C\}$$

where a_t is called alarm function and C threshold. The design of the alarm function defines different change-point detection methods while the values of the threshold reflects the degree of sensitivity of the stopping rule. The alarm function may be based on the likelihood ratio

$$L(s,t) = \frac{f_{\mathbf{Q}_t}(\boldsymbol{q}_t | \ \theta = s \leq t)}{f_{\mathbf{Q}_t}(\boldsymbol{q}_t | \ \theta > t)}$$

where $f_{\mathbf{Q}_t}(\boldsymbol{q}_t|A)$ is the conditional joint density function of the random variables $(Q_1, \dots, Q_t) = \mathbf{Q}_t$ given A and where $\boldsymbol{q}_t$ is the vector of the observed values of $\mathbf{Q}_t$. Assuming indpendence of the variables $Q_1, \dots, Q_t$ the likelihood ratio simplifies to

$$L(s,t) = \prod_{u=s}^{t} \frac{f_1(q_u)}{f_0(q_u)}$$

where $f_0(q_u)$ is the marginal conditional density function of Q_u given that the shift has not occurred by time u and $f_1(q_u)$ is the marginal conditional density function of Q_u given that the shift occurred in or before time u.

The conditional density function of the Q score at time t given that the data is encrypted (i.e. uniformly distributed) is

$$f_E(q_t) = \frac{q_t^{k/2-1} e^{-q_t/2}}{2^{k/2}\Gamma(k/2)}$$

where k is the number of degrees of freedom, i.e. the number of classes (which in this study is 8 as explained above).

For the non-encrypted files, the conditional density function of the Q score is modeled by αX where $X \in \chi^2(k)$ and $\alpha > 1$ supposedly reflecting the inflated deviances from the uniform distribution had the data been encrypted. Thus

$$f_{NE}(q_t) = \frac{\partial}{\partial q_t} \mathbb{P}(\alpha X < q_t \mid X \in \chi^2(k))$$

$$= \frac{\left(\frac{q_t}{\alpha}\right)^{k/2-1} e^{-q_t/2\alpha}}{\alpha 2^{k/2}\Gamma(k/2)}$$

is the density function of non-encrypted data Q score. This means that two cases of shift in didstribution are possible:

- Shift from non-encrypted to encrypted data in which case

$$L(s,t) = \prod_{u=s}^{t} \frac{f_E(q_u)}{f_{NE}(q_u)} = \alpha^{k(t-s+1)/2} \exp\left(-\frac{\alpha-1}{2\alpha}\sum_{u=s}^{t} q_u\right)$$

- Shift from encrypted to non-encrypted data in which case

$$L(s,t) = \prod_{u=s}^{t} \frac{f_{NE}(q_u)}{f_E(q_u)} = \alpha^{-k(t-s+1)/2} \exp\left(\frac{\alpha-1}{2\alpha}\sum_{u=s}^{t} q_u\right)$$

To detect whether the shift in distribution has occurred or not according to the stopping rule τ mentioned above, an alarm function should be specified. Two of the most common choices here are:

- CUSUM [24]: $a_t = \max_{1 \le s \le t} L(s,t)$,
- Shiryaev [25]: $a_t = \sum_{s=1}^{t} L(s,t)$.

Other possible choices are e.g. the Shewhart method, the Exponentially Weighted Moving Average (EWMA), the full Likelihood Ratio method (LR) and others, see e.g. [20] for a more extensive presentation of different methods.

For the CUSUM alarm function, as $\arg\max_{1 \le s \le t} L(s,t) = \arg\max_{1 \le s \le t} \ln L(s,t)$ the alarm function is simplified without any loss of generality by using the log likelihood values instead.

For both cases the alarm functions can be expressed recursively which facilitates collecting and treating the data as follows.

- The alarm function for shift from non-encrypted to encrypted data for the

- CUSUM method is

$$a_t = \begin{cases} 0 & \text{for } t = 0 \\ \max\left(a_{t-1} + \frac{k \ln \alpha}{2}, 0\right) + \frac{1-\alpha}{2\alpha} q_t & \text{for } t = 1, 2, 3, \ldots \end{cases}$$

- Shiryaev method is

$$a_t = \begin{cases} 0 & \text{for } t = 0 \\ \alpha^{k/2} e^{\frac{1-\alpha}{2\alpha} q_t} (1 + a_{t-1}) & \text{for } t = 1, 2, 3, \ldots \end{cases}$$

- The alarm function for shift from encrypted to non-encrypted data for the
- CUSUM method is

$$a_t = \begin{cases} 0 & \text{for } t = 0 \\ \max\left(a_{t-1} - \frac{k \ln(\alpha)}{2}, 0\right) + \frac{\alpha-1}{2\alpha} q_t & \text{for } t = 1, 2, 3, \ldots \end{cases}$$

- Shiryaev method is

$$a_t = \begin{cases} 0 & \text{for } t = 0 \\ \alpha^{-k/2} e^{\frac{\alpha-1}{2\alpha} q_t} (1 + a_{t-1}) & \text{for } t = 1, 2, 3, \ldots \end{cases}$$

3. Results

3.1 Evaluation

To quantify the quality of different methods, the performance is compared regarding relevant properties such as the time until a false alarm, delay of a motivated alarm, the credibility of an alarm and so on. The threshold is commonly calibrated against the Average Run Length ARL^0 which is defined as the expected time until an alarm when no parameter shift occurred (which means that this is a false alarm). It is crucial to have the right threshold values for the methods to perform as specified. Setting the threshold such that ARL^0 is 100, 500, 2500 and 10 000 respectively (the most common values here are $ARL^0 = 100$ and 500 but the higher values are also considered since the value of ARL^0 defines the number of clusters/time points that are treated before a false alarm and the shift could occur very far into the HDD) properties of the methods regarding delay and credibility of a motivated alarm can be compared. Of course, if the threshold is low, this will lead to more false alarms (detection of a change when there is none) but setting a too high threshold will lead to a drop of sensitivity of the method to detect a shift (higher delay between a shift and its detection) and consequently an increased probability of missing a real shift in distribution.

Usually the expected delay, $ED(\nu) = E_\nu(\tau - \theta \mid \theta < \tau)$ (expectation of the delay of a motivated alarm; see **Figure 1**) or Conditional Expected Delay $CED(t) = E(\tau - \theta \mid \tau > \theta = t)$ (expectation of the delay when the change point is fixed equal to t)

are very important because, for example in health care, the goal is to quickly detect problems to be able to save lifes (**Table 1**).

However, in the case of detecting encrypted code, expected delays are less relevant as a measure of performance since the data can be handled without any time aspect: the goal is to detect accurately where the encrypted data is located. A method with high expected or conditional expected delay merely means a slightly less efficient procedure.

A more relevant performance indicator, in this case, is for instance the predictive value $PV = P(\theta < \tau)$ (the probability that the method signals alarm when the change-point has occurred; see **Figure 5**) or the percentage of detected encrypted files that is discovered while running the process and how to improve it (see **Figure 2**).

While running the process, the method will stop at some time, τ, and then estimate the change point, θ, by maximizing the likelihood function by using the data after the last previous alarm and the newly detected change-point. This estimated change-point, $\hat{\theta}$, can be either before or after the true change-point θ. One could increase the intervals where encrypted data were discovered. This would lead to missing less encrypted data (see **Figure 2**) but also brute-forcing more non-encrypted data (**Table 2**).

Therefore the difference between the change-points and the alarms according to the method is calculated. Since the proportion of encrypted data relative to the total amount of data on the HDD is unknown, the expected proportion of error is suggested. This is to say, given two change-points, θ_1 and θ_2, and two stopping times, τ_1 and τ_2, the expected proportion of error is $E\left(\frac{|\tau_1-\theta_1|+|\tau_2-\theta_2|}{\theta_2-\theta_1}\right)$. However, this value has a sense only when there are no false alarms between τ_1 and τ_2.

If there are false alarms between τ_1 and τ_2, the proportion of undetected encrypted data was added to the proportion of error to determine the proportion of the error made relative to the size of the encrypted data. Assuming that there are n false alarms $\tau'_1 < \ldots < \tau'_n$ in $[\tau_1, \tau_2]$, called *expected inaccuracy*, or *EI* for short, is defined as follows.

$$EI = E\left(\frac{|\tau_1-\theta_1|+|\tau_2-\theta_2|}{\theta_2-\theta_1} + \sum_{i=1}^{n/2}\frac{\tau'_{2i}-\tau'_{2i-1}}{\theta_2-\theta_1}\right) \quad (1)$$

The *EI* was measured for different values of the parameter ν in the geometrical distribution of the change-points for different methods (see **Table 3** and **Figure 3**).

ν	**Methods**							
	CUSUM				**Shiryaev**			
	100	**500**	**2500**	**10,000**	**100**	**500**	**2500**	**10,000**
0.2	4.7844	7.1087	9.5520	11.6905	4.9672	7.3116	9.7634	11.9159
0.15	4.7674	7.0788	9.5109	11.6495	4.8933	7.2409	9.6760	11.8015
0.07	4.7278	7.0162	9.4401	11.5695	4.7455	7.0610	9.4786	11.6114
0.05	4.7176	7.0017	9.4224	11.5420	4.7021	6.9975	9.4308	11.5441
0.02	4.6957	6.9712	9.3860	11.4940	4.6422	6.9144	9.3175	11.4477
0.01	4.6870	6.9581	9.3698	11.4693	4.6150	6.8633	9.2743	11.3973

Table 1.
Values of expected delays ED for the CUSUM and Shiryaev methods for values of $ARL^\circ = 100, 500, 2500, 10000$ *for a shift from encrypted to compressed data.*

i	Percentage	
0	CUSUM	Shiryaev
1	0.960254	0.961280
2	0.971053	0.971274
3	0.976242	0.978665
4	0.979266	0.980872
5	0.983681	0.985597
6	0.986248	0.986101
7	0.990101	0.990101
8	0.993371	0.994931
9	0.994326	0.995682

Table 2.
Percentage of encrypted files that are detected when the interval of detected change points $[\tau_1, \tau_2]$ is arbitrary replaced by $[\tau_1 - i, \tau_2 + i]$. Typically, with large i, the values are very close but not exactly equal to 1: This happens because the change points are very close (i.e less than 10 clusters for example) and the method does not detect any change. Then no cyphertext is detected at all.

ν	Percentage	
	CUSUM	Shiryaev
0.20	0.113791	0.116934
0.15	0.112024	0.112757
0.07	0.095589	0.096884
0.05	0.086831	0.088897
0.02	0.062135	0.062066
0.01	0.044261	0.043975
0.005	0.030960	0.030376
0.001	0.016548	0.016028

Table 3.
Expected inaccuracy, EI [see Eq. (1)], for different values of the parameter ν and for the two methods. For the application of discriminating between encrypted and non-encrypted data on a hard drive this may be considered to be reflecting the degree of fragmentation of the disk; the less fragmentation the farther apart are the change-points and thus the smaller the ν value, and vice versa.

3.2 Complete procedure

The complete process is the method returning a segmentation separating suspiciously encrypted data and most likely non-encrypted data of an HDD; information to be further used to target the brute force cryptanalysis efficiently. This procedure runs a likelihood ratio based change-point detection method and as soon as it detects a change, calculates the maximum likelihood estimator of the change-point to determine where the change-point most likely is located. It will then start over from the location of this estimated change-point with the same method for online change-point

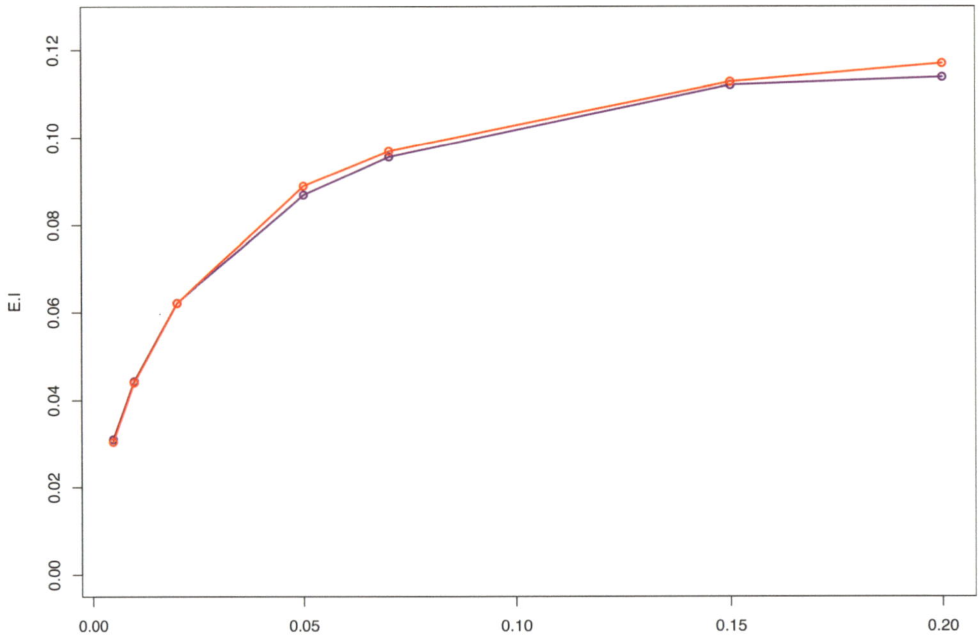

Figure 3.
Expected inaccuracy EI for the CUSUM (blue graph) and Shiryaev (red graph) procedure. One can see that the Shiryaev procedure is little less accurate when ν increases but slightly more accurate for small ν than the CUSUM procedure.

detection except that the likelihood ratio is reversed modifying the alarm function to fit the opposite change-point situation, and so on.

3.3 Thresholds and experimental values

The first step is to determine the thresholds rendering $ARL^0 = 100, 500, 2500$ and $10\,000$, for both the CUSUM and Shiryaev methods, for a shift from encrypted to non-encrypted and vice versa. The change-points are commonly geometrically distributed with parameter ν. Here the average time before a change-point is expected to be rather high (several hundred or thousand maybe) as the method deal with 64-byte clusters in an HDD of surely several hundreds of Giga or Tera Bytes. Thus, since $E(\tau) = 1/\nu$, the focus is on very small values of ν for the methods to be reasonably sensitive.

Commonly values of ARL^0 are 100 or 500 to make other properties relevant for comparisons. In the case of this application, however, large values of ARL^0 as 2500 and 10 000 are studied because the first change-point might not occur until far into the HDD. Adjusting the threshold by simulating data can take very long if ARL^0 is large (2500 or 10 000, especially for the Shiryaev method). In this case, it can take several hours or even up to days to compute the threshold. Therefore, having a way of predicting the threshold by extrapolation would be desired i.e. having an explicit relation between ARL^0 and the threshold C. Intuitively, if ARL^0 is larger, more data will be taken into account implying a threshold proportionally larger. Indeed, as ARL^0 increases more data is used in the procedure and the threshold is therefore proportionally increased from how much more data is treated in the procedure. In the CUSUM case, since the alarm function is defined as the log-likelihood ratio, this results in a threshold being a $\log ARL^0$ (**Figures 4** and 5; **Tables 4** and 5):

- for a shift from compressed data to encrypted data:

$$C = 0.965524 \cdot \ln\left(0.030655 \cdot ARL^0 + 0.494603\right)$$

- for a shift from encrypted data to compressed data:

$$C = 0.997767 \cdot \ln\left(0.912316 \cdot ARL^0 + 7.294950\right)$$

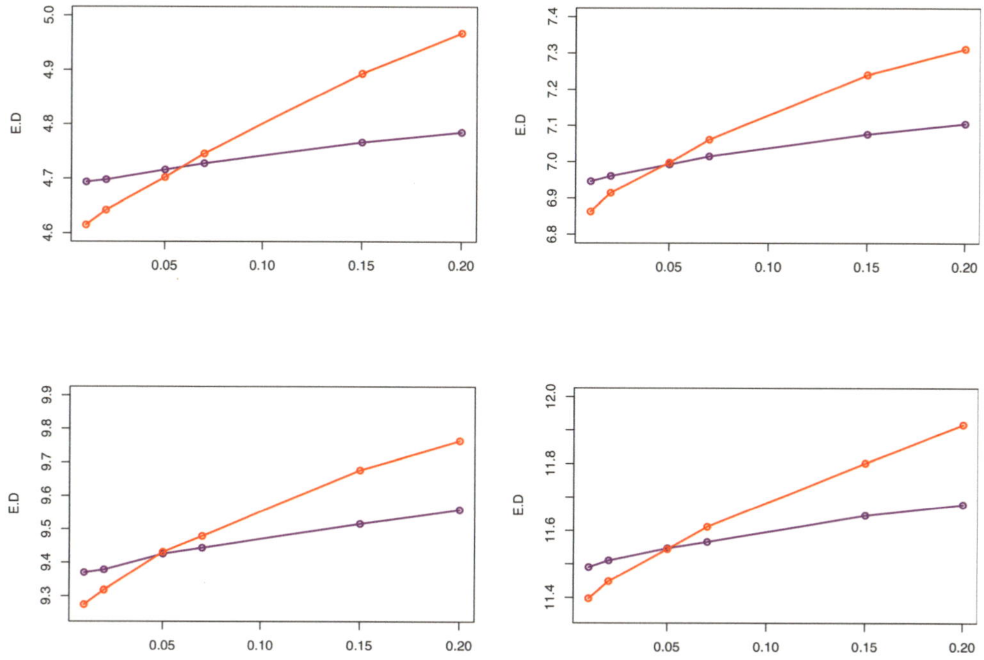

Figure 4.
Expected delays ED for a shift from encrypted to compressed data for the CUSUM procedure (blue) and Shiryaev procedure (red).

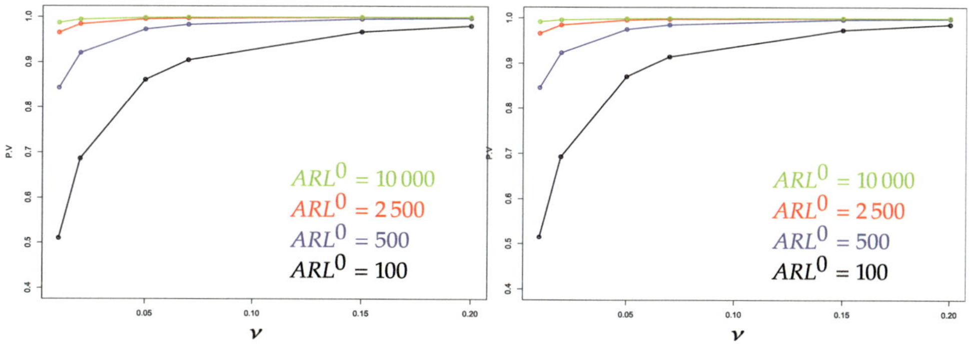

Figure 5.
Predictive values for a shift from compressed to encrypted data for the CUSUM procedure (left) and for the Shiryaev procedure (right).

ARL^0	Thresholds			
	CUSUM		Shiryaev	
	$NE \rightarrow E$	$E \rightarrow NE$	$NE \rightarrow E$	$E \rightarrow NE$
100	1.2260	4.5801	64.0313	44.1271
500	2.6529	6.1250	323.0625	221.8125
2500	4.2188	7.7120	1618.219	735.4088
10,000	5.5296	9.0990	6475.0547	4441.8413

Table 4.
Values of the thresholds for the CUSUM and Shiryaev methods for $ARL^{\circ} = 100, 500, 2\,500, 10\,000$ specified for detecting a shift from non-encrypted to encrypted data (indicated by NE $\rightarrow$ E for short) and for shift from encrypted to non-encrypted data (indicated by E $\rightarrow$ NE) respectively.

ν	Methods							
	CUSUM				Shiryaev			
	100	500	2500	10,000	100	500	2500	10,000
0.20	0.8950	0.9862	0.9984	0.9997	0.9859	0.9984	0.9998	1.0000
0.15	0.8727	0.9805	0.9974	0.9995	0.9736	0.9967	0.9996	0.9999
0.07	0.8031	0.9604	0.9933	0.9986	0.9147	0.9852	0.9976	0.9995
0.05	0.7634	0.9482	0.9907	0.9978	0.8708	0.9754	0.9957	0.9991
0.02	0.6171	0.8942	0.9784	0.9944	0.6927	0.9235	0.9847	0.9964
0.01	0.4712	0.8207	0.9590	0.9890	0.5153	0.8463	0.9661	0.9918

Table 5.
Predictive value $PV(\nu) = P_{\nu}(\theta < \tau)$, i.e. the probability that a shift has occurred when an alarm is signaled, for the CUSUM and the Shiryaev methods, for values of $ARL^{\circ} = 100, 500, 2\,500, 10\,000$ and with different values of the parameter ν in the geometric distribution of the change-points.

For Shiryaev, the threshold is a linear function of ARL^0:

- for a shift from compressed data to encrypted data:

$$C = 0.647578 \cdot ARL^0 - 0.726563$$

- for a shift from encrypted data to compressed data:

$$C = 0.444214 \cdot ARL^0 + 0.294281$$

4. Conclusion

Using the change-point statistical process, a method to detect encrypted data in HDD was successively designed. This method is using the fact that encrypted data is uniformly distributed as opposed to other types of files. The method was designed to detect even a change with the closest files to encrypted files which are compressed data. As this method even detects a small change in the data, any bigger change will be even easier detected. Therefore this process is likely to detect encrypted data among any type of data.

Quick and accurate detection of a change is commonly the desired property of change-point detection methods. In many applications such as medicine, finance, environmental science etc., time aspects of the methods are a matter of interest, e.g. expected delay in detection of a shift or probability of detecting a shift within a specified time interval. Here, however, this time aspect is not of primary interest since the data remain the same during the whole process. Here the need is to detect correctly recognized encrypted data. Therefore the probability of correctly detecting encrypted data is more relevant here. This probability shows that the method detects more than 96% of the encrypted data which is good and by extending the intervals, the method detects more than 99% of the encrypted data. By assuming that the change-points are not too close—which is a plausible assumption since it is unlikely that files are so small if the device is not too fragmented—then the method, by adding a little margin to the intervals, quickly detects 100% of the encrypted data.

The Shiryaev method turns out to be slightly better in the more important respects compared to the CUSUM method. Although the expected delay ED is bigger than CUSUM for large values of the parameter ν in the distribution of the change points, it is smaller for small values of ν which is the most relevant case for detecting encrypted data in an HDD. The Shiryaev method also detects more encrypted data than the CUSUM method and has a slightly higher predictive value PV.

All in all, this means that both methods designed with the suggested modeling, perform very well with a slight preference to the Shiryaev method for detecting encrypted data in an HDD.

To summarize, a thorough comparison between the proposed method and the aforementioned methods [3, 4, 8, 9, 18] for the situation with streamed data would be the obvious next step in this research. Also other methods, potentially building on the Kolmogorov–Smirnov statistic or the Shannon entropy and by using other anomaly detection of machine learning could be interesting candidates in such a race.

Acknowledgements

The authors wish to express their gratitude to Mattias Weckstén at Halmstad for good ideas and previous readings of the manuscript University and to Linus Nissi (previously Linus Barkman) at the Police Department of Southern Sweden for earlier work in the area.

Conflict of interest

The authors declare no conflict of interest.

Nomenclature

ARL^0	average runlength in control
ARL^1	average runlength out of control

ED	expected delay of motivated alarm
CED	conditionala expected delay of motivated alarm given that the change occurred at a specified time-point
PV	predictive value

Author details

Eric Järpe[1*] and Quentin Gouchet[2]

1 Halmstad University, Halmstad, Sweden

2 Atsec Information Security, Austin, Texas, USA

*Address all correspondence to: eric.jarpe@hh.se

References

[1] Malhotra P. Detection of encrypted streams for egress monitoring [thesis]. UMI Microform 1447482. Ames, Iowa: Iowa State University; 2007

[2] Razali NM, Wah YB. Power comparisons of Shapiro-Wilk, Kolmogorov-Smirnov, Lilliefors and Anderson-Darling tests. Journal of Statistical Modeling and Analytics. 2011; **2**(1):21-33. Available from: https://www.researchgate.net/publication/267205556_Power_Comparisons_of_Shapiro-Wilk_Kolmogorov-Smirnov_Lilliefors_and_Anderson-Darling_Tests [Accessed: December 29, 2021]

[3] Hahn, D, Apthorpe, N and Feamster, N. Detecting Compressed Cleartext Traffic from Consumer Internet of Things Devices [Internet]. 2018. Available from: https://arxiv.org/abs/1805.02722 [Accessed: December 29, 2021]

[4] Choudhury P, Kumar KR, Nandi S, Athithan G. An empirical approach towards characterization of encrypted and unencrypted VoIP traffic. Multimedia Tools and Applications. 2020;**79**:603-631. DOI: 10.1007/s11042-019-08088-w

[5] Bassham LE, Rukhin A, Soto J, Nechvatal J, Smid M, Barker E, et al. A Statistical Test Suite for Random and Pseudorandom Number Generators for Cryptographic Applications, National Institute of Standards and Technology, Special Publication 800–22 [Internet]. 2010. Available from: https://tsapps.nist.gov/publication/get_pdf.cfm?pub_id=906762 [Accessed: December 29, 2021]

[6] Walker J. Pseudorandom number sequence test program, Fourmilab.ch [Internet]. 2008. Available from: https://www.fourmilab.ch/random [Accessed: December 29, 2021]

[7] Kozachok AV, Spirin AA. Model of pseudo-random sequences generated by encryption and compression algorithms. Programming and Computer Software. 2021;**47**(4):249-260. DOI: 10.1134/S0361768821040058

[8] Yao Z, Ge J, Wu Y, Lin X, He R, Ma Y. Encrypted traffic classification based on Gaussian mixture models and hidden Markov models. Journal of Network and Computer Applications. 2021;**166**:1-13. DOI: 10.1016/j.jnca.2020.102711

[9] Li Y, Lu Y. ETCC: Encrypted two-label classification using CNN. Security and Communication Networks. 2021; **2021**:1-11. DOI: 10.1155/2021/6633250

[10] Andersson F, Nelander Hedqvist K, Ring J, Skarp A. It-inslag i brottsligheten och rättsväsendets förmå ga att hantera dem. Brottsförebyggande rå det, report. 2016;**17**:1-152. Available from: https://bra.se/publikationer/arkiv/publikationer/2016-09-30-it-inslag-i-brottsligheten-och-rattsvasendets-formaga-att-hantera-dem.html [Accessed: December 29, 2021]

[11] Swedish Civil Contingencies Agency. Informationssäkerhet – trender 2015, Myndigheten för Samhällsskydd och Beredskap [Internet]. 2015. Available from: https://www.msb.se/sv/publikationer/informationssakerhet–trender-2015 [Accessed: December 29, 2021]

[12] The Swedish Police. Polisens rapport om organiserad brottslighet 2015, National operations depertment [Internet]. 2015. Available from: https://docplayer.se/844230-Polisens-rapport-om-organiserad-brottslighet-2015-polismyndigheten-nationella-operativa-

avdelningen-maj-2015.html [Accessed: December 29, 2021]

[13] Bischoff, P. Best Disk Encryption Software – the Top 5 Tools to Secure Your Data, Comparitech, May 13 [Internet]. 2020. Avaialble from: https://www.comparitech.com/blog/information-security/disk-encryption-software [Accessed: December 29, 2021]

[14] Paninski L. Estimating entropy on *m* bins given fewer than *m* samples. IEEE Transactions on Information Theory. 2004;**50**(9):2200-2203. DOI: 10.1109/TIT.2004.833360

[15] Barkman L. Detektering av krypterade filer [thesis]. diva2:428544. Halmstad, Sweden: Halmstad University; 2011

[16] Westfeld A, Pfitzmann A. Attacks on steganographic systems-breaking the steganographic utilities ezstego, Jsteg, Steganos, and S-Tool and some lessons learned. In: Information Hiding, Third International Workshop IH'99, September–October Dresden. Germany: Springer; 2000. pp. 61-76

[17] Pearson K. On the criterion that a given system of deviations from the probable in the case of a correlated system of variables is such that it can be reasonably supposed to have arisen from random sampling. Philosophical Magazine. 1900;**50**(302):157-175. DOI: 10.1080/14786440009463897

[18] Casino F, Choo KKR, Patsakis C. HEDGE: Efficient traffic classification of encrypted and compressed packets. IEEE Transactions on Information Forensics and Security. 2019;**14**(11): 2916-2926. DOI: 10.1109/TIFS.2019.2911156

[19] Frisén M. Properties and use of the Shewhart method and its followers. Sequential Analysis. 2007;**26**(2):171-193. DOI: 10.1080/07474940701247164

[20] Frisén M. Statistical surveillance. Optimality and methods. International Statistical Review. 2003;**71**(2):403-434. DOI: 10.1111/j.1751-5823.2003.tb00205.x

[21] Frisén M, de Maré J. Optimal surveillance. Biometrika. 1991;**78**(2): 271-280. DOI: 10.2307/2337252

[22] Järpe E, Wessman P. Some power aspects of methods for detecting different shifts in the mean. Communications in Statistics, Computation and Simulation. 2000; **29**(2):633-646. DOI: 10.1080/03610910008813632

[23] Järpe E. Surveillance, environmental. In: El-Shaarawi AH, Piegorsch WA, editors. Encyclopedia of Environmetrics. 2nd ed. Chichester: Wiley; 2013. pp. 2150-2153. DOI: 10.1002/9780470057339.vas065.pub2

[24] Page ES. Continuous inspection schemes. Biometrika. 1954;**41**(1/2): 100-115. DOI: 10.1093/biomet/41.1-2.100

[25] Shiryaev AN. On optimum methods in quickest detection problems. Theory of Probability and Its Applications. 1963; **8**(1):22-46. DOI: 10.1137/1108002